**Cambridge Studies in Applied Ecology and Resource Management**

The rationale underlying much recent ecological research has been the necessity to understand the dynamics of species and ecosystems in order to predict and minimize the possible consequences of human activities. As the social and economic pressures for development rise, such studies become increasingly relevant, and ecological considerations have come to play a more important role in the management of natural resources. The objective of this series is to demonstrate how ecological research should be applied in the formation of rational management programmes for natural resources, particularly where social, economic or conservation issues are involved. The subject matter will range from single species where conservation or commercial considerations are important to whole ecosystems where massive perturbations like hydro-electric schemes or changes in land use are proposed. The prime criterion for inclusion will be the relevance of the ecological research to elucidate specific, clearly defined management problems, particularly where development programmes generate problems of incompatibility between conservation and commercial interests.

*Timber production and biodiversity conservation in tropical rain forests*

**Also in the series**

Graeme Caughley, Neil Shephard & Jeff Short (eds.)   *Kangaroos: their ecology and management in the sheep rangelands of Australia*

P. Howell, M. Lock & S. Cobb (eds.)   *The Jonglei canal: impact and opportunity*

Robert J. Judson, K. R. Drew & L. M. Baskin (eds.)   *Wildlife production systems: economic utilization of wild ungulates*

M. S. Boyce   *The Jackson elk herd: intensive wildlife management in North America*

Mark R. Stanley Price   *Animal re-introductions: the Arabian Oryx in Oman*

R. Sukumar   *The Asian Elephant: ecology and management*

K. Homewood & W. A. Rodgers   *Maasailand ecology: pastorialist development and wildlife conservation at Ngorongoro, Tanzania*

D. Pimentel (ed.)   *World soil erosion and conservation*

R. J. Scholes & B. H. Walker   *An African savanna: synthesis of the Nysvley study*

T. D. Smith   *Scaling fisheries: the science of measuring the effects of fishing 1855–1955*

Jim Hone   *Analysis of vertebrate pest control*

# TIMBER PRODUCTION AND BIODIVERSITY CONSERVATION IN TROPICAL RAIN FORESTS

**Andrew Grieser Johns**

*Research Associate Oxford Forestry Institute*

CAMBRIDGE
UNIVERSITY PRESS

PUBLISHED BY THE PRESS SYNDICATE OF THE UNIVERSITY OF CAMBRIDGE
The Pitt Building, Trumpington Street, Cambridge, United Kingdom

CAMBRIDGE UNIVERSITY PRESS
The Edinburgh Building, Cambridge CB2 2RU, UK
40 West 20th Street, New York NY 10011–4211, USA
477 Williamstown Road, Port Melbourne, VIC 3207, Australia
Ruiz de Alarcón 13, 28014 Madrid, Spain
Dock House, The Waterfront, Cape Town 8001, South Africa

http://www.cambridge.org

First published 1997
First paperback edition 2004

Typeset in 11/14pt Times

*A catalogue record for this book is available from the British Library*

ISBN 0 521 57282 7 hardback
ISBN 0 521 60762 0 paperback

Dedicated to the memory of
**ARTHUR MORRELL**

# CONTENTS

# FOREWORD

It is a great privilege to be asked to write the foreword to this book for three reasons.

First, the scientific quality of the book and its professionally significant implications are outstanding and the author should be congratulated. The roles of forests in human welfare and environmental sustainability are now widely recognized, including the role of man-made plantations. Throughout the world there is an increasing recognition of the needs for and benefits of trees and forests for an ever-widening range of products and services. Among the products, timber is still the most important because of its wide range of uses and because of the increasing sophistication of the wood-using industries in converting timber efficiently, making long-lasting products and reducing adverse environmental impacts. Unfortunately, the timber industry has received considerable ill-founded criticism in the last decade concerning its alleged role in causing tropical deforestation. Equally there has been criticism of plantations, particularly with exotic species, on the alleged grounds that they damage soils, use all available agricultural water, reduce natural biodiversity and deprive poor people of either land rights or employment. While there are undoubtedly cases where such criticisms are justified, there is increasing concern throughout the industry to seek wiser management and sustainable exploitation of forests for all their benefits. This book should add to our understanding and achievement of such aims.

Second, it is a privilege to recognize the contribution of Mr Daniel Kemp, Chairman of Timbmet Group Limited, who by financial contribution to the Oxford Forestry Institute facilitated the preparation of this book. Mr Kemp came to Britain in 1938 as a refugee and helped his father establish a timber business. Following his father's death in 1959 he developed this into a thriving timber importing business that is widely reputed throughout the tropical world as an honest and concerned company prepared to discuss frankly forest management as well as wood use and the company's environmental impact. Mr Kemp was a great friend of Mr Arthur Morrell to whose memory he wishes this book to be dedicated.

Third, it is my privilege to share in the recognition of Arthur Morrell and I am grateful to Mr Bob Plumptre and Mr Geoff Pleydell for their help in preparing this summary of Arthur's life.

Arthur Morrell spent all his working life in the British timber trade. He started in 1939 at a very young age as a junior member of the staff of the Nottingham branch of Fitchett and Woollacott; he became a director of the same firm in 1966 and joint managing director in 1980. In the early 1980s he moved to run the operations of Parker Kislingbury in Herefordshire and later became Chairman of the hardwood interests of Mallinson-Denny, which was subsequently purchased by Hunter Timber within the Wickes Group. Retirement in 1990 was only nominal. He became Chairman of the National Hardwood Association of the Timber Trade Federation, a leading player in the industry's 'Forests For Ever' programme and the trade adviser to the UK delegation to the International Tropical Timber Organization (ITTO). He was also a consultant to Timbmet and a non-executive director of his old company, Fitchett and Woollacott.

It is hardly surprising that, after 51 years in the hardwood timber trade, Arthur knew something about timber and the trade. He had extensive knowledge, not only of the commonly traded timbers, but had engaged in repeated trials of 'lesser-known' species from as early as the 1960s. It might have been expected that after such a long career in the trade he would be somewhat intolerant of strong, anti-trade, environmental views, particularly those which ignored the practicalities of marketing and utilizing timber. However, it was clear that his huge love for wood and his knowledge of its versatility as a material made him particularly determined that wood should be conserved and used in ways that promoted the continued existence and good management of forests and all the benefits accruing from them.

He was a practical conservationist in the best sense and cared enough to be prepared to spend an immense amount of time and energy in trying to bridge the gap between the environmental lobby and the trade. This care was demonstrated by the fact that he was always ready to listen to anyone's views, however extreme, answering them fairly, but very often openly and honestly on what he considered were their merits.

His skills as a chairman and as an arbitrator in a dispute were remarkable, partly because of his respect for people in their own right, which brought with it an ability to see both sides of a dispute, and partly because his sense of proportion and humour made it difficult to be angry with him for any length of time. This was demonstrated many times in ITTO meetings during long and sometimes heated discussions on the problems of harvesting and managing tropical forests sustainably. In his work with ITTO he frequently promoted or supported the funding of pragmatic and achievable projects designed to solve these problems.

Perhaps the most enduring memory of Arthur is of a person for whom nothing, however apparently trivial it might be, was too much trouble for him, even when he was under considerable pressure, providing it was going to help someone.

Jeffery Burley
Director, Oxford Forestry Institute

# PREFACE

In the late 1970s, I was somewhat surprised to find myself conducting the first detailed work on the ecological effects of timber logging in a tropical rain forest. Surprised, that is, that so little had been conducted beforehand. Due mainly to the medium of television, enormous public interest had been kindled concerning the fate of the rain forests. In response to this interest, a flood of popular and scientific books and articles reporting on the consequences of tropical forest loss had been unleashed. The extent of clearance of tropical forest for shifting or plantation agriculture, cattle ranching or other human activities was widely and often acrimoniously debated.

All the fuss focussed exclusively on deforestation, which could conveniently be tracked by satellite technology and did not require much by way of actual field research. After deforestation, timber logging was the second most important human influence on tropical forests. It was known to affect 50 000 km$^2$ of rain forest annually during the late 1970s and 1980s. Yet no-one had considered it appropriate to go into the field to research the effects logging might have on rain forest biodiversity. Forests subjected to timber logging were invariably included within 'deforestation' statistics. They were considered as lost to wildlife and ignored in forest conservation strategies.

Look down from an airplane flying over any extant tropical forest and the differences are quite obvious. Areas at the edge of the forest are continually being cleared to support the expanding human populations. These appear as bare ground, irregular outlines of crop fields or ranches, or serried ranks of plantation crops. Although a transition may be gradual, at some point the overt evidence of human presence will give way to an irregular tree canopy with occasional cleared patches and a snaking road network. This is the forest that has been logged but nevertheless remains as forest. Normally, quite a lot of this will need to be traversed before reaching forests that appear pristine. If, indeed, there are any.

Travel the same distance on the ground and the differences are even more obvious. Agricultural areas, villages and an abundance of people give way to stands of trees, dense regenerating growth and an absence of people. Signs of animal life begin at the forest edge and become commoner in the interior. One thing becomes clear very quickly, the assumption that logged forests support none of the biodiversity of unlogged forest is plain wrong.

In 1983 I approached Cambridge University Press with the idea of publishing a book pointing out the possibilities for the integration of tropical forestry and biodiversity conservation strategies. At that time this was considered almost heretical. It remains controversial in some quarters, although it is now a much more acceptable approach. In 1983 my ideas were barely formed, however, and relied heavily on data I had just collected in South-East Asia. The idea was nevertheless encouraged by Robin Pellew, then an editor at the Press, although the importance of a global approach was stressed. I subsequently went back into the field to conduct further research, first to South America and later to Africa. By 1995 sufficient information had been collected, by myself, co-workers and independent scientists throughout the tropics, that a global analysis could confidently be undertaken.

The book relies a great deal on my original fieldwork, since this forms a large part of the available information from tropical forests. I would thus like to acknowledge the financial support received at various stages over the last 18 years from the following: the National Environmental Research Council, the Overseas Development Administration and The Royal Society of the UK, the US Government (National Institutes of Health) and USAID, WWF-International, WWF-US, WWF-Malaysia and the Wildlife Conservation Society. My thanks also for the use of field facilities run by the Forest Research Institute of Malaysia, the Sabah Foundation and Makerere University.

Over the years many people have assisted the development of research by offering comments on manuscripts, through discussion, or through provision of unpublished data or reports of limited distribution. In this context I would particularly thank the following: C. Fairgreave (Edinburgh), M. Heydon (Fordingbridge), L. Holbech (Copenhagen), P. Howard (Accra), M. Kalyakin (Moscow), J. Karr (Blacksburg, VA), A. Kemp (Pretoria), F. Lambert (Bangkok), R. Lowe (Ibadan), L. Roche (Dublin), T. T. Struhsaker (Durham, NC), S. Sutton (Manchester), J-M. Thiollay (Paris), L. White (Edinburgh) and T. Whitmore (Cambridge). Key figures who have assisted the development of research over the years are D. Chivers (Cambridge), C. Marsh (Vientiane), A. Marshall (Aberdeen) and R. Mittermeier (Washington, DC).

I am particularly grateful to the following who have read and commented on sections of this book: N. Brown (Oxford), C. Dranzoa (Kampala), D. Earl (Kingsbridge), T. Johns (Bracknell), A. Plumptre (Budongo Forest Project), D. Pomeroy (Kampala), R. Putman (Oban) and V. Reynolds (Oxford). J. Burley (Oxford) and B. Plumptre (Oxford) kindly read through the entire manuscript prior to submission. I am also grateful to J. Burley for contributing the preface.

The following acknowledgements are made for the inclusion of previously published illustrations:

M. Heydon/Journal of Applied Biology, Blackwell Science Ltd. (Fig. 3.1).
L. Bruijnzeel (Fig. 3.2).
D. Nicholson (Fig. 4.2).
M. Heydon (Fig. 4.4).
Biological Conservation, Elsevier Science Ltd. (Fig. 8.6).
J. Hendrison/Wageningen Agricultural University (Figs. 9.2 and 9.3a).

Finally, my thanks to my sons, Micha and Sasha, without whom the book would have been finished much sooner but without whom life would be less enjoyable. And to my wife, Bettina, for every kind of support.

The completion of the book was assisted by a memorial award from the Oxford Forestry Institute, donated by Mr Daniel Kemp (Timbmet Group Limited) in memory of the late Mr Arthur Morrell. That commercial interests should participate in the publication of work on tropical forest conservation is an encouraging indication of how the two often opposing camps are converging.

Andrew Grieser Johns
Brackley, Northamptonshire

# EXPLANATORY NOTE

A logged forest is a transient ecosystem moving to regain the structure that existed prior to logging. The most important factors affecting the extent to which it differs from primary forest are the time that has elapsed since the logging event and the degree of damage caused by the logging operation. In the text of this book these data are included for all references to a logged$^{(12:50)}$ forest. This example indicates that data refer to a forest logged 12 years previously at which time 50% of trees were destroyed. Elapsed periods date from the onset of logging rather than its completion.

On occasion, published studies fail to provide these data. The information may sometimes be inferred from other published material from the same location. At other times, an estimate can be made based on other information given (such as timber volumes removed) and comparative damage levels in other locations within the same geographical region. Where data are not verified, they are preceded by a question mark, as in logged$^{(12:? < 35)}$ forest. An absence of any figures, as in logged$^{(12:?)}$ forest, indicates no estimate was considered appropriate.

# 1

---

# The issues

## Introduction
### The importance of the tropical timber trade

At the beginning of the 1990s, global consumption of wood products was running at around 3400 million m$^3$ annually. By the end of the decade, this is expected to reach 4200 million m$^3$ (World Bank 1991). About 55% of the total demand is for fuelwood, primarily in developing countries where in the early 1990s some 500 million people were living in or near forests. The remaining 45% is composed of industrial forest products, two-thirds pulp and one-third timber.

Hardwood products from natural tropical forests (excluding plantations of tropical species) accounted for only around 16% of the worlds timber trade in the early 1990s (Barbier *et al.* 1994) and only 3% of wood products traded in the UK. However, hardwoods represented 70% of industrial wood exports from tropical countries (tropical forests produced only 7% of the worlds woodpulp).

In the early 1990s timber exports were worth US$10 000 million annually to tropical countries (up from US$600 million annually in the early 1970s). This ranks fifth out of non-oil commodities (twice the value of rubber and three times the value of cocoa). During the early 1990s, between 20 and 30% of the total originated in three countries: Indonesia, Malaysia and Brazil (production from the Philippines, formerly a top exporter, has declined drastically since the mid-1980s).

The importance of timber to local economies is considerable. At the end of the 1980s the forestry sector in developing countries made an annual contribution of around US$35 000 million to gross domestic product (GDP). In Malaysia and Indonesia this amounted to 3–6% of the national GDP, but in the Malaysian state of Sabah forestry provided 70% of total Government revenue. Malaysia, and particularly Sabah, were able to realize a high percentage rent capture through effective taxation systems. However, timber frequently fails to achieve more than a small percentage of its potential contribution to Government revenue through problems of setting and collecting appropriate rents (Repetto 1988, Barbier *et al.* 1994). The potential contribution to developing countries' economies is much higher.

In addition to providing Government revenue directly, forestry can be a major employer. For example, in the early 1990s the forestry sector was employing about 151 000 people in Malaysia and 28% of the entire workforce of Gabon.

### The regional status of tropical forests

There are two broad divisions of tropical forest: tropical moist forest and tropical dry forest. Tropical moist forests themselves are normally divided into tropical rain forest, which makes up about 97% of the total, tropical monsoon (seasonal) forest and tropical thorn forests. The global distribution of the forest types is determined mainly by climate. Tropical rain forests are widely distributed through the equatorial belt. Monsoon forests spread from India through South-East Asia to Melanesia. Thorn forests are rather limited in their distribution, occurring in significant quantities only in India.

The tropical moist forests may be separated into a number of biogeographical regions (Figure 1.1). The limits of these regions may be determined by physical barriers, such as the Andean mountains in South America, by climatic factors causing interspersion of dry forests or arid vegetation, or by historical

Fig. 1.1. Major tropical forest regions (recent historical limits of tropical forest). Numbers refer to Table 1.1.

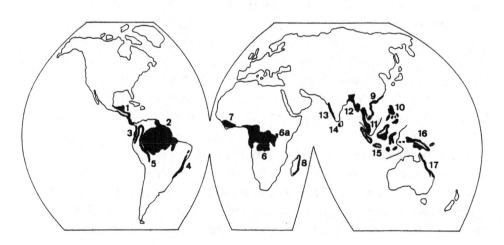

factors such as the meeting points of Asian- and Australasian-dominated biotas in central Indonesia.

The recent status of tropical moist forests is summarized in Table 1.1, together with representative levels of vertebrate species endemism. The importance of tropical forests in preserving biodiversity is emphasized. For example, in the New Guinean region (Irian Jaya, Papua New Guinea and southern Melanesia), the remaining 780 000 km$^2$ of tropical forests (7.2% of the world total) support 22% of the world's endemic tropical forest birds or 5% of all bird species.

According to the statistics in Table 1.1, tropical moist forests (unlogged and logged) covered an estimated 10.9 million km$^2$ in 1990. This amounted to 52% of its original (post-glacial) area. The most reliable estimates suggest that 12.6 million km$^2$ of tropical forest existed at the beginning of the 1980s, or 61% of the original area (FAO 1988; World Bank 1991). The proportion of the total that was undisturbed by agriculture or timber industries fell from about 71% of the remaining forests in 1980–85 to 60% in 1990 (World Bank 1991).

Rates of loss or conversion of tropical forest are notoriously difficult to estimate accurately, and are susceptible to manipulation (Whitmore & Sayer 1992). However, the most objective estimates suggest that an average of 90 000 km$^2$ of previously undisturbed forest became affected each year during the decade 1971–80, and 169 000 km$^2$ each year during the decade 1981–90 (FAO 1991a). This is in agreement with the above estimate of 1.7 million km$^2$ of forest lost between 1980–90.

During the 1980s, around 50 000 km$^2$ of forest were logged for timber each year (Sayer & Whitmore 1991): the remainder was converted to other forms of land use.[1] Much smaller areas were reforested through natural or assisted regeneration. Total forest areas continue to decline.

There is considerable geographical variation in rates of deforestation (Table 1.2). The greatest loss of undisturbed forest has been in tropical South-East Asia which also has the highest percentage of logged forest. The percentage share of logged forest continues to increase in South-East Asia. The decreasing percentage of logged forest in tropical Africa reflects the exhaustion of resources in most of West Africa and the loss of logged forest area to shifting cultivation.

Country by country statistics detailing forest loss and disturbance rates are available for the 1970s and 1980s (FAO 1981, 1988, 1991a). There is great variation in loss rates which is usually a reflection on human population

---

[1] Of the 50 000 km$^2$ of forest logged annually, a small but unknown proportion consisted of forest that had already been logged and was thus already disturbed. The total area converted to non-forest was therefore somewhat larger than 119 000 km$^2$ annually.

pressures. For example, Bangladesh and most West African countries are around 90% deforested while 90% of Amazonian *terra firme* forest remains largely undisturbed (although pressures are mounting). As a general rule, pressures faced by forests are inversely related to their remaining area, except where major transmigration schemes are planned (e.g. Sumatra and Kalimantan in Indonesia, the state of Rondônia in Brazil).

Table 1.1. *Status of the worlds major tropical rain forest regions in 1990, with representative species endemism*

| Region | Recent historical total area (km² × 10³) | Area remaining (km² × 10³) | Preserved forest area (km² × 10³) | Production forest area (km² × 10³) | No. endemic birds | No. endemic primates |
|---|---|---|---|---|---|---|
| 1. Central American | 335 | 217 | 43 | 135 | 228 | 3 |
| 2. Amazonian | | | | | | |
|    i. flooded[a] | 70 | 60 | 3 | 30 | 64 | 3 |
|    ii. terra firme | 550 | 4969 | 353 | 448 | 430 | 30 |
| 3. Colombia coastal | 117 | 96 | 11 | 20 | 79 | 3 |
| 4. Atlantic coastal | 1205 | 146 | 31 | 40 | 171 | 10 |
| 5. Andean | 505 | 371 | 86 | 75 | 251 | 2 |
| 6. Zairoise | 5275 | 1955 | 116 | 1086 | 159 | 21 |
| 6a. Albertine rift | 126 | 48 | 2 | 31 | 30 | 1 |
| 7. Guinean | 2121 | 112 | 13 | 80 | 25 | 4 |
| 8. Malagasy | 103 | 42 | 4 | 4 | 30 | 11 |
| 9. Indochinese[b] | 675 | 319 | 29 | 76 | 14 | 3 |
| 10. Philippine | 240 | 67 | 6 | 61 | 68 | 1 |
| 11. Malayan | 230 | 162 | 29 | 118 | 6 | 2 |
| 12. Myanmar-Bengal[b] | 920 | 504 | 8 | 262 | 7 | 3 |
| 13. South Indian[b] | 350 | 39 | 15 | 16 | 15 | 2 |
| 14. Sri Lankan[b] | 22 | 17 | 5 | 9 | 11 | 1 |
| 15. Indonesian | 2106 | 958 | 139 | 547 | 204 | 18 |
| 16. New Guinean | 800 | 780 | 52 | 327 | 509 | — |
| 17. North Australian | 13 | 11 | 2 | 1 | 13 | — |
| Totals | 20 721 | 10 873 | 949 | 3366 | 2314 | 116 |

There is a high level of accuracy for figures from some regions (such as north Australia), but many are approximations only (particularly those from Indochina, Central and South America).

Estimates are by region rather than country, thus eastern Malaysia is included within the Indonesian region, Irian Jaya is included within the New Guinean region, etc.

Some countries have little or no formally declared production forest estate: for these the estimate is of the area under logging concessions.

Endemic species are defined as those that occur only in tropical high forest and not in adjacent habitats. Forest endemics occurring in more than one biogeographical region are also excluded.

[a]Flooded (periodically inundated) forests are composed of igapó forest and várzea, the latter of which is part forest and part swamp.

[b]Includes rain forest and monsoon forest.

*Principal sources*: Sayer *et al.* (1992); Collins *et al.* (1991); WRI (1994); (numerous other sources consulted for individual country statistics and for lists of endemic species).

## Problems facing tropical forest management

From the mid-1980s, imports (mostly wood-based products such as panels, pulp and paper) into tropical countries began to exceed the value of exports (mostly timber) by 30–50%. It is expected that by the year 2000 less than 10 tropical countries will be net exporters (which will not include former major exporting countries such as Nigeria and the Philippines, both of whose forest resources are critically depleted). Although many tropical countries still possess extensive tracts of forest, or soil and climatic conditions amenable to the establishment of industrial hardwood plantations, there has been little economic investment in maintaining or upgrading the resource.

There has been a failure across the board in systems of tropical high forest management so far devised (Mergen & Vincent 1987). This is despite the fact that at least some systems are biologically sound and should succeed in practice (Whitmore 1991). Reasons for their failure are complex, but relate primarily to economic, social and political factors. Some principal problems are outlined below.

Table 1.2. *The changing distribution of undisturbed and logged forests between broad geographical regions*

| | Neotropics | Tropical Africa | Tropical Asia & Australasia |
|---|---|---|---|
| Total area of undisturbed tropical moist forest (%)[a] | | | |
| 2000 years ago | 57 | 18 | 25 |
| 1980 | 68 | 18[d] | 14 |
| 1990 | 70 | 17[d] | 13 |
| Total area of logged tropical moist forest (%)[b] | | | |
| 1980 | 28 | 23 | 49 |
| 1990 | 28 | 19 | 53 |
| Mean country deforestation rate during 1980s (%)[c] | 1.4 | 2.1 | 0.9 |

[a]Preserved forest area plus undisturbed forest outside of the preserved areas.
[b]Includes the logged portion of the production forest estate (the percentage of reserved production forest that has actually been logged varies between regions) plus ungazetted logged forest.
[c]Calculated from figures for representative countries given in Whitmore & Sayer (1992).
[d]Less than 1% in the Guinean region.
*Sources*: Brown & Lugo (1984); WRI (1994).

*Insufficient reinvestment in the resource*

Concession agreements issued to timber companies generally require them to undertake certain management activities during the life of the agreement. However, concession agreements almost never cover a full harvesting cycle and thus offer little incentive for investment in the long-term productivity of the forest (Barbier 1993). In dipterocarp forests of South-East Asia, for example, concessions are usually issued for 21–25 years although polycyclic logging systems have a minimum harvesting cycle of 35–40 years and monocyclic systems rather longer (Poore *et al.* 1989). The size of issued logging concessions is also frequently less than that required to make a viable long-term management unit.

It is, of course, possible to create an economic climate for investment in the resource. The necessary changes in current taxes and fee structures that would ensure reforestation by the private sector without a major loss of profits have been demonstrated for the Côte d'Ivoire (Mendoza & Ayemou 1992). Requirements for replanting of forests heavily damaged during logging are commonly stated in the terms of the concession, but problems arise in enforcing such rules. It is sometimes less expensive for concessionaires to ignore such regulations and pay resulting fines.

In addition to a failure to restore timber crops in managed forest, there has been a general lack of reforestation or afforestation of degraded or vacant land. Reforestation of areas excessively damaged by past logging or shifting agriculture is expensive, but may be economically viable under certain conditions (Korpelainen *et al.* 1995). Establishment of industrial plantations has always been a neglected option as it generally produces a less valuable product at a higher cost than harvesting of natural forest. During the early 1980s, less than 6000 km² of industrial plantations (including softwoods) were being established per year in the developing countries (WRI 1985). Planting had increased to 30 000 km² per year by the early 1990s, but this was due mostly to an increase in tree planting through community forestry and fuelwood programmes (Ball 1992). In the early 1990s only 350 000–439 000 km² of industrial plantations existed in the tropics and subtropics combined (World Bank 1991, FAO 1992). However, the potential for industrial plantations to lessen pressures on the natural forest resource is growing (see Chapter 8).

*Inadequate infrastructures*

There has long been a shortage of trained foresters, insufficient financial investment in silvicultural research and inadequate administrative structuring in most tropical countries. The first is a particular problem in implementing improved forest management systems in tropical countries

(Poore *et al.* 1989), and particularly so with respect to measures for biodiversity conservation for which no formal training programmes exist outside of Australia.

Many forestry operations fail to recycle profits in the industry, particularly offshore-based companies which still dominate the industry in many tropical countries. Greater commitment on the part of governments is also required towards maintaining productive forests. In the past, forest departments in developing countries have been regarded primarily as generators of central revenue: their activities are underfunded, with profits from forest fees being absorbed directly by national treasuries (Palmer & Synnott 1992).

### Forest invasion

During the 1980s between 300 and 400 million people moved into tropical forest lands, doubling the population of these areas.

Logged forest, with its network of new roads, is susceptible to encroachment by slash-and-burn agriculturalists, or, in the Neotropics, by cattle ranchers. In many African countries more than 50% of logged forest area is subsequently deforested while there is virtually no deforestation of previously unlogged forest (Barbier *et al.* 1994). In Ecuador, the concession system for timber companies was abandoned following the failure of the Government to prevent deforestation of logged forest areas (Palmer & Synnott 1992).

In addition to clearing areas for cultivation of crops, colonists degrade remaining forest areas by cutting of pole-sized trees for construction or fuelwood (these trees would otherwise form the next timber crop). This has been a main cause of the failure of forest management in some areas, such as coastal Colombia. In eastern Amazonia, it has been demonstrated that fires set for clearance of land for agriculture or cattle ranches often spread into logged or degraded forests, causing severe damage (Uhl & Buschbacher 1985).

There is no doubt that timber harvesting is, unfortunately, a major indirect cause of deforestation. This seriously undermines forest management planning in many countries. Governments are often unable or unwilling to enforce their own regulations against encroachment into state-controlled forests. The problem is difficult to overcome since it arises from deep-rooted social problems, notably the unequal distribution of land ownership (Ayres 1986a). Three possible solutions are to increase governmental capacity to manage the resource, to privatize the resource or to devolve management responsibilities onto local populations. The first requires financial aid of the order of US$300–1500 million annually (Barbier *et al.* 1994), and the second has been shown generally to be ineffective. Experiments are underway in many tropical forest regions to examine the potential of the third option.

### Inefficient species utilization

Most logging operations are highly selective: only a few species of tree are targeted. Damage levels can be high and wastage considerable as many of the trees knocked down during road building and harvesting activities are not marketable. Furthermore, it may be necessary to increase the increment of the valuable target species in regenerating forest to improve the projected regeneration cycle, and interventions are both expensive and prone to failure. This has led in the past to foresters recommending the abandonment of tropical forest management altogether and replacing the species-rich forests with industrial plantations (some have later had second thoughts: Leslie 1987).

An approach to the problem has been to develop a market for a greater number of tree species, enabling more intensive use of forest and less wastage. This has been successful only in supplying domestic markets through a domestic processing capacity (e.g. in Australia). There is a great deal of inertia in the international marketplace. Importing countries will generally not consider lesser-known species (LKSs) as long as supplies of preferred timbers are maintained (Freezaillah 1984). Increasing the prices of preferred tropical species is not reflected in shifts to the use of cheaper tropical LKSs (Vincent *et al.* 1990). In some countries there is evidence of substitution of temperate timber (Vincent *et al.* 1991). In the European construction industry plywood can be substituted with solid synthetic panels (Barbier *et al.* 1994). Unless market conditions change, LKSs can be marketed only as chips, which are of low economic value and can have high social and environmental costs (Lamb 1990).

### Timber pricing systems

Timber tax structures are generally imbalanced and are inefficient in generating revenue, both because rents are set too low and because actual rents captured may only be a small proportion of rents due. In Cameroon during the late 1980s, forest fees and stumpage charges per $m^3$ of hardwood represented only 2–4% of the export value of logs. Other African countries, such as Guinea, had forest fees set at less than 1% of the export value of the timber at this time. In addition to low forest fees, it was estimated that most West and Central African countries had a very low fee collection rate: only about 20% in Congo during the late 1980s (Barbier *et al.* 1994). A variety of techniques exist for avoiding forest fees, such as illegal (undeclared) logging, under-measurement of processed products, misdeclaration of species, and financial instruments such as transfer pricing schemes (Barbier *et al.* 1994).

Attempts have been made by many Governments to encourage processing industries producing value-added products by keeping the prices of raw timber products artificially low. This can have two major disadvantages. First, it can lead to over-production and wasteful uses of resources. In Indonesia, preferential tax structures which encouraged processing and export of plywood led to an expansion of highly inefficient plywood mills which provided little additional Government revenue while causing far more environmental damage than if logs had been exported to more efficient plywood mills elsewhere (Gillis 1988). Second, it provides no incentive for investment in maintaining the forests through careful harvesting practices or replanting. Stumpage prices should have a crucial role in facilitating the transition of the forest sector from its current dependence on old-growth forest to second-growth forests and in coordinating processing capacity with timber stocks.

Governments have failed to avoid allowing excessive profit margins for concessionaires, regardless of their management practices, which has encouraged excessive exploitation of the forest resources. Profits need to be tied in some way to the standards of practice. Tax structures do not currently favour observance of best-practice forestry (Barbier *et al.* 1994).

It has been suggested that sustainable forest management might be encouraged by appropriate subsidies. Such a scheme could be financed in two ways. First, by tax transfers, where stumpage and other charges in producing countries are increased and importation taxes in consuming countries decreased (OFI 1991). Second, by revenue transfers, where a proportion of import taxes collected by consumers would be returned to the producers (Barbier *et al.* 1994). The rationale for provision of a trade subsidy on timber from well-managed forests rests on the suppositions that sustainable management can be achieved only at increased costs and that prices paid by the consumer are not high enough for sufficient returns to filter down to the concessionaire's end of the chain. Neither of these is necessarily true. Furthermore, increasing revenue capture by the Government does not provide incentives for better management by individual concessionaires, indeed it may encourage poor management practices.

An alternative approach would be the introduction of a tiered system for marketing higher-priced timber from concessions where best practice is observed. This may be viable if consumers are prepared to pay a premium for products from well-managed forests and if this is realized in prices paid to the concessionaire. Surveys conducted in the UK during the late 1980s and early 1990s suggest that 25–65% of people may be willing to pay more for 'certified' timber products, and that a surcharge of 6–10% is generally acceptable (Barbier *et al.* 1993). A survey by WWF (1991) gave an average acceptable surcharge of 13.6%, but this is likely to vary with the prevalent economic climate. Timber products from certified sources, be they companies or whole countries, are likely to be of rising importance in the timber trade in the next decades. However, it remains to be seen if higher prices for certified timber will force further substitution of non-tropical timber products.

### Political instability

Forestry, with its complex administrative structures and long management cycles is particularly susceptible to disruption through macro-economic and political instability. On the economic side, hyperinflation of the kind common in developing countries can quickly influence the cost-effectiveness of management operations. On the political side, unrest and civil conflicts can prevent management programmes being carried out, or timber concessions can be awarded on terms highly favourable to the contractor (which usually means fewer rules protecting the resource) in return for political support. In many countries, such as the Philippines, politicians are prominent on the board of logging companies and it is generally they rather than foresters who determine concession policies (Poore *et al.* 1989). Factors of this kind can disrupt even the most carefully planned management operations on local and national scales.

A further disruptive element, as far as forestry is concerned, is the increasing pressure on the industry from environmental protection movements, normally based in developed countries. Changing attitudes in consumer countries are reflected in governmental decision-making and policies of their lending agencies. In Queensland, forestry operations were halted amid considerable controversy when a sizeable proportion of the forest estate became a World Heritage Site in 1988 (Moulds 1988). Many Dutch Government-funded forestry research projects, such as the CSS initiative in Suriname, have been affected by cuts in funds available for tropical forestry. In Uganda, sizeable portions of the commercial forest estate were re-gazetted as National Parks during the early 1990s, largely as a result of loan conditions imposed by the US Government. The extent to which forest preservation is a more efficient means of biodiversity conservation than multiple-use management depends on local attitudes and on the financial and other resources available to the custodian government departments (Tabor *et al.* 1990).

## Problems facing tropical forest preservation

In the worst case scenario, neither the existing system of National Parks and other pristine protected areas, nor any conceivable such system, will be sufficient to preserve forest biodiversity. It is generally agreed that this state of affairs has already been reached in the temperate forests of the United States (Franklin 1989, Alaric Sample *et al.* 1993). It can equally be assumed to have been reached in some tropical forest biogeographical regions.

In 1990, 8.7% of current tropical forest area was legally protected in National Parks and equivalent reserves (Table 1.1). The total protected area continues to increase at a slow rate. As a percentage of the remaining undisturbed forest the rate of increase is more rapid, due to the shrinkage of undisturbed forest areas. The creation of new protected areas is also geographically imbalanced: most new reserves are being created in the largest remaining forest areas, where they are least urgent. Furthermore, the amount of preserved land that is actually physically protected against intrusion by hunters, settlers or refugees, or excluded from oil and mineral prospecting, hydroelectric schemes and trunk road developments, is actually much less. In some parts of the world, legally protected areas exist on maps but there is little or no enforcement on the ground.

In tropical countries, more than 70% of resident vertebrate species are commonly reported to be dependent upon closed (undisturbed and lightly disturbed) forest. The corresponding figure for invertebrates is hard to arrive at: actual diversity, life histories and the degree of persistence in secondary forest are all largely unknown (Stork 1988, Sutton & Collins 1991). The degree of endemism, both of vertebrates and invertebrates, is very high in tropical forest regions (Gentry 1986).

There has been much debate as to how many tropical species are likely to be lost if deforestation rates continue (summarized in Reid 1992, Heywood & Stuart 1992). In general, even those extinction rates predicted from objective mathematical models tend not to be realized, and there is little field evidence for large scale extinctions brought about by deforestation (Dodson & Gentry 1991). Nevertheless, even the loss of a few species reduces biodiversity and there are few data concerning the erosion of genetic diversity within species caused by habitat fragmentation (UNEP 1995).

### Forest and biodiversity conservation: the rationale for an integrated approach

In the early 1980s the area of selectively logged tropical forest was reported to exceed that of preserved and protected undisturbed forest by a ratio of 4:1 (Brown & Lugo 1984). By the early 1990s this had reached a ratio of 8–10:1 (World Bank 1991). These figures are difficult to verify, however. First, an unknown but sizeable portion of the global protected forest area (949 000 km² in 1990: Table 1.1) is logged or otherwise disturbed. Second, a usually undefined portion of the production forest estate (3 366 000 km² in 1990: Table 1.1) has not yet been logged and part of the area is likely to remain unlogged, for various reasons. Third, much logged forest exists outside of production forest estate or formally declared timber concession areas (e.g. in

1990 4 148 000 km² of Amazonian forest was gazetted neither as protected nor as production forest and some of this had been or was being logged).

However the ratio is constructed, the disparity between protected primary forest and logged forest area will continue to increase. Although many logged areas are subsequently clear-felled for small-scale or plantation agriculture, or for forestry plantations, large areas of so far undisturbed forest continue to be logged each year. No new primary forest is being created, of course.

Two basic problems have faced forest and wildlife conservation in the tropics: a shortage of undisturbed areas and erosion of what little has been set aside. In the first case, financial considerations and land-use trends often preclude the preservation of large areas of economically non-productive forest. Even where ecotourism or other non-destructive uses are envisaged, these areas can still be financial sinks (Dixon & Sherman 1990, Lindberg & Huber 1993). In the second case, few reserves can be considered inviolate, particularly in regions with rapidly expanding rural populations and where participatory management of the forest resources is envisaged.

Economic justification for retention of large tropical forest areas can occasionally be provided on the basis of sustainable annual harvesting of non-wood products, fruits and so on (e.g. Peters *et al.* 1989). In rare cases, economic cost-benefit analysis has indicated that a combination of non-destructive in-forest activities and improved out-forest activities can be more profitable than forest harvesting or clearance (e.g. the much-cited case of the Korup National Park in Cameroon: Ruitenbeck 1988). Timber production from these particular forests is generally marginally economic.

Although ecotourism can generate comparable revenue to forest rents in small areas (a few tens of km²), this form of forest use is generally not economically justifiable on a large scale, especially if forest rents are increased to a more realistic percentage of timber value. Wild game harvesting from primary forest can be economically important (Caldecott 1988), as can the production of butterflies in some cases (Oldfield 1988), but revenues accruing to the government are unlikely to equal forest rents and these alternative uses should be possible in managed production forests if adequately regulated. Environmental values of intact forest (watershed protection, carbon fixation, etc.) need not be compromised by the minimal alteration of the vegetation arising from responsible management. Other reasons for preserving large intact forest areas (preservation of biodiversity for its own sake, possible threats to resources of possible pharmaceutical value, etc.) cannot be fitted into an economic equation at all, although they may have moral value.

Given the high rates of return that can be achieved from timber harvesting, it is not surprising that alternative forms of use are under-developed. Although often poorly regulated, timber extraction is the most cost-effective use of almost all tropical forest areas. Development of regional biodiversity conservation strategies need to take this into account: to an increasing extent tropical forest species will have to be managed within the sphere of economic production from those forests. Forest managed for sustainable timber production should be an integral part of conservation strategies (Johns 1985, Sayer 1991).

In the past, disturbed forests have generally only been considered in regional or local conservation strategies where intact forest is drastically reduced or its standing timber is of premium value (e.g. Davies & Payne 1982). If considered only at this stage, the value of the disturbed forest for species conservation may be compromised by poor felling practices, or invasion by hunters or agricultural colonists. It is clearly more useful to integrate conservation procedures within forest planning and development.

The tropical timber industry faces a number of problems of its own, and these have to be appreciated in planning the development of joint management strategies. The tendency has been to concentrate on sustained production of timber, not necessarily sustaining the forest ecosystem (Noss 1993). The two goals of sustainable timber production and sustainability of the forest ecosystem are not necessarily incompatible; indeed they may be interlinked.

### The concept of sustainable management

At the 1992 Oxford Conference on Tropical Forests, Synnott (1992) underlined the real problem by posing the following riddle:

> What is it that very few people have seen, but which many people have defined, that in practice barely exists, or that doesn't exist at all (according to some people), that is impossible anyway (according to other people), but which, if ITTO, TTF and WWF have their way, will become one of the dominant forms of land use in the tropics by the year 2000?

Tropical forestry has always been problematic and sustainable forest management even more so. The majority of current forest management systems operate close to the borders of profitability. It is a general truism that management systems designed to safeguard the resource are less profitable than indiscriminate creaming off of valuable timber from easily accessible forests (especially where forest rents and other fees can be circumnavigated, and no reinvestment in the resource undertaken). Protective management systems also tend to be less profitable than other forms of land use.

The result has been that many countries which researched and applied appropriate forest management systems have still failed to protect the resource, either because of deforestation for other land uses or because foresters themselves have lost control over the resource. For example, the Malayan uniform system has been considered quite successful in lowland dipterocarp forest in peninsular Malaysia, but these forests have been almost completely replaced by agricultural plantations and the system does not work well in the hill dipterocarp forests where conditions for silviculture are rather different (Mok 1992). Very little of the worlds tropical forest can be considered effectively managed. In the late 1980s only 10 000 km$^2$ of tropical forest was considered to meet identified criteria for sustainable management (Poore *et al.* 1989), although much larger areas would require little additional effort to meet these criteria (Palmer & Synnott 1992).

The essence of forest management is that the forests remain productive of timber, and other resources, indefinitely. Harvesting regimes applied should be sustainable. Sustainable management may be defined by a forester as:

> To harvest forest in such a way that provides a regular yield of forest produce without destroying or radically altering the composition and structure of the forest as a whole. (Wyatt-Smith 1987)

The conservation community, which has taken the idea of sustainable management very much on board in recent years, has rephrased the definition somewhat:

> [Sustainable forestry] must meet the economic needs of the industry while the loss of biodiversity and forest functioning associated with logging must be acceptable. (Robinson 1993)

As pointed out in an earlier paper (A. Johns 1992), all definitions of sustainability include an indeterminate qualifier. What does a forester regard as 'radical' and what level of change is 'acceptable' to the conservationist?

Beyond the basic definitions, the divergence in opinion as to what constitutes sustainable management becomes increasingly apparent. The forester concentrates on sustainable timber production:

> Thus it is more than maintaining the rate of production at the low level of natural forest, which merely replenishes natural mortality. It is used in the sense of enhanced production through silvicultural practices, while conserving the protective role of the vegetation and the genetic pool of all species other than those regarded as weed species which compete with, and suppress, favoured timber trees. (Wyatt-Smith 1987)

This makes the point that the economics of timber management frequently require some degree of alteration of the natural system towards a higher density of timber trees. A managed forest will not necessarily maintain the balance of species found in undisturbed forest. In broad terms, however, the natural regeneration potential of tropical forest sets limits on the activities of foresters as minimal intervention is of paramount importance to the long-term economics of timber production.

The conservation viewpoint tends to centre on the Leopoldian ideal: maximizing flexibility by preserving all components of a system (Leopold 1949). The conservationist thus concentrates on preservation of the ecosystem as a whole:

> The point therefore is not that a tropical forest cannot be sustainably used – any intensity of use is potentially sustainable. The more intense the human use of a forest, however, the greater will be the loss of biological diversity. (Robinson 1993)

Robinson (1993) assumes a linear relationship between disturbance and loss of biodiversity. The relationship may actually be much more complex (see Chapter 7). Under natural regeneration (minimal intervention) systems most if not all plant and animal species may be expected to persist in the long term, providing the forest management units are large, whereas at certain levels of disturbance the resilience of the ecosystem is overcome and large species losses begin to occur.

The presence of some animals in the regenerating forest will be of benefit to foresters (pollinators and dispersers of tree seeds, predators of tree pests, etc.) and it will be to the foresters' advantage to maintain them. On the other hand, some animal species may retard the regrowth of commercial tree seedlings and conflicts of interest may arise. Some herbivorous mammals are both seed dispersers and destructive browsers. There may well be points of conflict. Nevertheless, there is little doubt that forests managed for timber production can form an important supplement to totally protected areas and may even provide the best chance of maintaining many tropical forest species.

A so far undefined balance between the economics of timber production and requirements for the preservation of biodiversity is the basis for sustainable management of the tropical forests. As has been recognized in principle since the beginning of the 1980s (IUCN/UNEP/WWF 1980), and more recently reinforced (IUCN/UNEP/WWF 1991), the development of forest management strategies should include an element of biodiversity conservation. What has been less frequently recognized is that the development of biodiversity conservation strategies should take the opportunities presented by the forest industrial sector into account.

The aim of this book is to examine the role of production forests in conserving tropical forest biodiversity, albeit in a somewhat altered state. Animal species or groups of species may or may not be able to co-exist alongside various forms of forest management. Where certain species are unable to co-exist, possibilities may exist for improving or adjusting management practices without detracting from the cost-effectiveness of the operation. Where the economic costs of providing for the conservation of certain species in production forests are too high, priorities for forest preservation in totally protected areas are more clearly defined.

### Summary

Tropical timber is an important commodity on the international market and the trade is vital to the economies of many tropical countries. Unfortunately, poor management of the resource has led to its being severely depleted in many biogeographical regions. Governments have failed adequately to protect the production forest estate and much has been converted, legally or illegally, into agricultural land. The continual loss of forest estate has severely hampered the ability of foresters to research and implement appropriate management strategies.

Deforestation has become a principal threat to tropical forest biodiversity. The forests of few biogeographical regions are adequately represented in well-protected National Parks or equivalent areas and demands are growing for preserved areas to prove themselves economically viable. The growing trend towards participatory resource management and the development of multiple-use options for protected areas suggest a growing realization that exclusive preservation of forest is an outdated concept and an unaffordable luxury.

Forestry is frequently the most cost-effective form of land use in many tropical regions, particularly hilly areas unsuited to plantation agriculture. Although poorly managed to date, forestry has the potential to provide an important economic resource while at the same time conserving tropical forest in a close to natural state. The integration of conservation activities into production forest management is becoming essential for the retention of tropical forest biodiversity in the long term.

# 2

## The history and development of tropical forestry

### Historical background
*The pre-colonial era*

In his *Kritias* Plato delivered a severe warning of the consequences of continued deforestation in upper Attica. No-one took much notice, with the result that twentieth century Greek foresters are still trying to reforest unproductive land which was degraded as long ago as the third century BC. The importance of maintaining forests is commonly realized only in retrospect.

It is commonly accepted that people moved into tropical forests only comparatively recently: perhaps as long as 40 000 years ago in South-East Asia and the Pacific, but only 10 000 years ago in the Neotropics and 3000 years ago in Africa. The descendants of these people are today among the oldest cultural groups on Earth (e.g. the Onge and Jarawa tribes of the South Andaman Islands, who still lack the knowledge to make fire). Many have complex relationships with the tropical forest based on sustainable use of its various resources and made possible by their very low population densities.

Tree-felling dates back to the advent of flint axes. Throughout history trees have been selected for qualities suitable to construction of houses, canoes or cultural artifacts, or for medicinal properties of their wood. Commercial harvesting dates back at least 2000 years, when the cedars of Cyprus and Lebanon were the first targets of the emerging timber industries of ancient Europe (Meiggs 1982). (Even by this time the cedars had been heavily felled: Noah's ark is recorded as having been built of Cyprus timbers, and Solomon's temple lined with boards from Lebanon.)

Tropical forest silvicultural practices also date back about 2000 years in Java and southern China. Tree species of particular cultural or religious significance were protected by legal or religious restrictions in several emerging civilizations. In Myanmar (formerly Burma), teak *Tectona grandis* was decreed a 'royal tree', all specimens belonged to the King. In Sri Lanka, satinwood *Chloroxylon swietenia* was regarded as an 'unlucky' tree, the timber of which could be used only in the construction of temples or royal dwellings.

The earliest commercial felling of tropical forest trees for export began around the seventh century in the coastal forests of East Africa. The Arab trading empire was expanding at this time and a few species of trees were selectively cut to supply the resulting boat-building industry.

As European empires became established, rare and valuable trees began to be extracted in large numbers from a variety of forest regions. Brasilwood *Caesalpinia echinata* and mahogany *Swietenia macrophylla* were cut in the Amazon basin for the manufacture of dyes and furniture, respectively. *Palaquium gutta* from Malaysia was used to produce gutta percha (a latex used as an insulator, primarily in submarine telegraph cables) which fetched a price of US$500 per pikul (60 grams) at the end of the nineteenth century. In Indonesia, the conifer *Agathis* was over-exploited in the production of Manila copal, a resin popularly used in the nineteenth century, mainly for varnishes.

### The colonial and post-colonial eras

The concept of felling and cutting large-sized timber for export dates mostly from the colonial era. Forest Departments were established in the early nineteenth century in British-ruled India and Myanmar, primarily concerned with safeguarding the teak resource. Teak was, at that time, highly important in ship-building and essential for the continued expansion of the British navy. Logs were sawn on site and extracted by animal (oxen and elephants) and water transport. Forest Departments followed in other Asian colonies and in African colonies by the early twentieth century.

By this time it was becoming economically feasible to ship timber or timber products to the burgeoning mass-production industries supplying the growing markets of European nations. The availability of home-grown hardwoods had declined drastically as the potential market increased considerably. Attention began to shift from the heavy, hard and naturally durable timbers that were the first to be exported towards medium and light hardwoods, many of which could be chemically treated for durability, using newly discovered processes, if this was desired in the end use. The tropical forests produced large, tall and straight-grained timbers idea for industrial and construction uses, and later plywoods and veneers. The timber industry concentrated on cutting the largest trees and Forest Departments set minimum girth limits accordingly.

Until the 1940s, trees were felled using axes and handsaws, and logs were hauled by animal teams to the nearest river for transport to coastal loading points. Only areas of forest close to navigable rivers could be logged, and damage levels were fairly low. After the Second World War, tropical forestry was revolutionized by the introduction of power saws and hauling and transport machinery. This opened up inter-fluvial areas to logging and changed the typical logging system from highly selective to highly intensive.

The actual products traded have changed markedly since the 1940s. Few countries now permit the export of round logs. Value-added products are the most common export item. Most tropical countries encourage domestic processing, which also generates revenue and employment on a local basis. This tends to be supported by lending to the forestry sector (World Bank 1991) as does reinvestment in the natural forest resource and establishment of hardwood plantations. Even countries which allow some round log exports require a minimum level of local processing (e.g. 75% in Gabon). This limits purely exploitative logging, at least to some extent.

### Tropical forestry practices

Tropical forestry can take a number of forms most of which concentrate on the production of timber and most of which involve the removal of selected species rather than the clear felling of whole stands. The predominance of selective felling differs from temperate forestry operations, where clear felling is the rule. Improving technology means that some tropical forests can now be cleared for the production of woodchips, but such operations are uncommon at present. Mixed laun (dipterocarp) chips from Indonesia may command only 30% of the price of eucalyptus chips, and their production is marginally economic even where 70% of the stand is utilized (especially where costs of reforestation are considered). The Amazonian forests, which are the largest remaining tropical forest areas but the least productive of commercial timber (Whitmore & Silva 1990), produce poor quality, dense, heavy woodchips and this is not likely to become an alternative use for the forests.

The export of timber is basically an exercise in transport at minimum cost, since the product is large and heavy, and somewhat under-valued on the international markets. The distances over which it is possible to transport timber vary according to its realizable value. In Amazonia, up to 60 species may be cut in the eastern basin close to the domestic markets of southern Brazil, but only two or three highly valuable species are cut in the isolated western forests, either for use in floating houses or boat construction, or for export. On a global scale, the major sources of tropical hardwoods have, until comparatively recently, been West African and South-East Asian nations which have developed extensive infrastructures and which lie on the main established trade routes. Global patterns have changed as most traditional source areas have suffered critical declines in timber stocks and economics have favoured the opening up of new supplies.

### Non-timber harvesting

Today, at the end of the twentieth century, tropical forestry is largely synonymous with the timber industry. This is a comparatively recent phenomenon. Timber accounted for only 55% of forest products traded from Indonesia in 1938 (although this had risen to 95% by the late 1980s, with most of the remaining 5% being made up of rattans). In some regions, non-timber products are still significant.

Harvesting of non-timber products from tropical forest is normally carried out on a small scale (although its potential may be otherwise: Fearnside 1989). In Amazonia, latex from *Castilloa ulei* and non-elastic gums from *Couma macrocarpa* and *Ragala sanguinolenta* are the basis for small but important local industries. A very few species of tree are enormously valuable as their woods possess qualities promoted by perfume manufacturers (e.g. sandalwood, *Santalum album*, from southern Asia). Brasilwood *Caesalpinia echinata*, from the Amazon region, has been widely exploited to supply the manufacturers of fine red and purple dyes, and has been eradicated from most accessible forest areas. An interesting case is *Aquillaria malaccensis* from Thailand, whose heartwood is sometimes infected by a black fungus which is much sought after in the perfume trade. All trees of this species are routinely felled in the hope that some may contain this fungus (which cannot be seen from the tree's exterior).

Although products from trees and other plants of tropical forests are widely used in traditional medicine, relatively few have achieved international significance. Most of these products can be more effectively harvested from plantings, such as the Amazonian climber *Chondrodendron tomentosum* and Zairean climber *Strophanthus gratus* which can be used to produce muscle relaxants used in heart medicine, and the pantropical yam *Dioscorea* which provides diosgenin, the steroid precursor of cortisone and oral contraceptives. In West Africa, the bark of *Pygeum* (*Prunus*) *africana* is widely exploited for the production of a drug for the treatment of prostatitis. Although plantations of this tree can be established quite easily, and are encouraged by many community forestry programmes, destructive use of wild stocks has led to the eradication of the tree from some areas, such as the montane forests of Cameroon. In Australia, chemicals contained in the seeds of the Moreton Bay chestnut *Castanospermum australe* have been identified as of potential use in combatting AIDS (Whitmore 1990).

The only non-wood products harvested on a large-scale are South-East Asian rattans and Amazonian palm hearts. Rattans, climbing palms of the genus *Calamus*, are harvested almost entirely from natural forest, although experimental plantations are being established (Dransfield 1988). Some 150 000 tonnes were traded annually in the late 1980s, worth US$2400 million per year. Rattans are often harvested alongside timber logging operations and their collection is usually poorly regulated. Edible palm hearts, from *Euterpe oleraceae*, *E. precatoria* and *Guilelma* spp., are currently obtained by near clear-felling of swamp and várzea in the upper Amazon where these trees can make up 60% or more of the stand. The exploitation of palm hearts has also been poorly regulated. The resource has been almost completely lost from the Atlantic forests of Brazil where *E. edulis* was the original mainstay of the trade (Galetti & Chivers 1995).

### Non-intensive timber harvesting

Most inhabitants of tropical forest fell trees for domestic uses, such as building houses or canoes, or for making vessels for the production of alcoholic beverages. Canoe construction is of particular interest, since it requires trees of very specific timber qualities. Communities along the Tana River delta in Kenya use only six species of tree, each with characteristic preparation times and useful lives on the river (Marsh *et al.* 1987). The most commonly cut tree, *Ficus sycamorus*, was exploited at the rate of one large tree per 1.5 km gallery forest per year, a rate of use that was rapidly depleting the resource. In the várzea forests of the upper Amazon, Ayres & Johns (1987) record eight species or species groups of tree being used to build canoes, but a distinct preference for a few species with a short preparation time relative to their working life. Again, the rate of use of these species was depleting the available stocks.

In East African coastal forests *Brachystegia speciformis* and *Brachylaena hutchinsii* are extensively felled for carving tourist souvenirs. Since small-sized trees can be used in this trade, and enforcement of minimum girth limits on exploited trees is lacking, regeneration of these trees is severely compromised. Similarly in Malaysia, *Intsia palembanica* is traditionally used for spirit carvings by forest dwelling *Orang Asli* tribes. Although spirit carving is a dying art, extensive marketing of imitations as souvenirs could result in cutting of small-sized trees in some forest areas, and this is one of the most valuable timber trees in South-East Asia.

In some parts of the tropics, timber is still extracted by hand using cross-cut saws and manual or water-borne transport. This form of harvesting does not produce an export-quality product and is essentially a means of supplying local markets. Nevertheless, it can be an important local industry.

In the upper Amazon várzea and igapó forests, villagers from the adjacent terra firme select and fell trees during the low-water season, leaving them on the ground until the forest floods. Logs can then be floated to a collection point, roped together into rafts and towed to a local sawmill, all without a need for expensive machinery (Ayres & Johns 1987). Only a few of the most valuable species are cut, although if these are of non-floating species they need to be roped to other logs for buoyancy. Rafting distances are kept as low as possible since towing logs requires capital input in the form of diesel fuel for the tug. Overall, profits are low and the level of damage to the forests is low (although desired timber species can become eradicated quite quickly).

By contrast, forest remnants close to urban or industrial centres can be quite lucratively harvested by hand. In Uganda, where the forest resource has in the past been poorly managed relative to market demand, illegal pit-sawing has been a major factor in resource depletion (Dranzoa & Johns 1992). In general terms however, pit-sawing is less damaging to the forest than mechanized logging, and it may have a role in community forestry, buffer zone management and the supply of domestic construction materials (Struhsaker 1987).

### Intensive timber harvesting

In a typical case, the development of tropical forest timber resources begins with traditional use, then expands to commercial use of a few species, then further expands to intensive use of many species. For example, in Queensland only *Toona australis* was used by the early settlers, 10 species were used by 1900, 30 species by 1930 and over 100 species by 1945. This is a function of the growing scarcity of the most sought-after species and thus the greater acceptance of alternatives (once their end uses are explored), and the growing acceptance of small-size species. In most parts of the world, the level of intensive use has already been reached.

Intensive timber harvesting operations are generally characterized by the use of heavy machinery. In a few areas, particularly the teak forests of upper Myanmar, elephants and oxen teams are still extensively used to handle logs, although the size of logs they can skid is limited. In the main producing countries of South-East Asia and Africa, animal skidding is rarely used. Instead a range of skidding, lifting and transport machines are employed, varying from the basic San Tai Wong truck of the Philippines and Malaysia, which has been in use since the early 1960s, to modified bulldozers, purpose-built tracked and wheeled skidders and loaders, and large transport vehicles (Jonsson & Lindgren 1990). Most logging operations require extensive road construction, although in mangrove and peat swamp forests of Malaysia extraction canals are excavated instead (Khan 1995).

As readily accessible lowland forests have been depleted, additional technology has been introduced for logging highland areas. A variety of cable systems have been used to haul logs up or down slopes, the most sophisticated of which are forms of high-lead yarding and skyline systems, originally developed in New Zealand and used with some success in the north-western USA. There has been some concern over levels of environmental damage caused by these systems, however. Aerial cable systems may reduce soil compaction on slopes and subsequent erosion, and may offer better protection for the residual seedling stock (Nicholson 1963). On the other hand, large areas of forest around landing sites are denuded of almost all vegetation cover, overall tree loss can be higher than under conventional logging systems (Blanche 1978, Johns 1989a), and regeneration of landing sites can take a very long time (Meijer 1970).

Logging with helicopters, tethered balloons and airships, which are environmentally the best option in steep forest, have been tried in a few temperate forest areas (Jonsson & Lindgren 1990). The last of these is probably the most cost-effective, but the technology is currently borderline and many operating problems have yet to be overcome. These systems are only likely to become feasible in extracting low volumes of extremely valuable product from otherwise inaccessible terrain.

### Forest management systems

Tropical forest management systems derive from the German tradition of forestry in Europe and began to be developed in India during the nineteenth century. The British took the resultant ideas to Africa around the beginning of the twentieth century: systematic research was underway in Nigeria by the 1920s. Silviculture began to be practised in South-East Asia at around the same time, but did not begin in Amazonia until the 1960s.

In some countries, such as Congo, a 'log and leave' strategy still predominates. Allowable diameter specimens of a few valuable trees are extracted based on a preliminary inventory. After logging the area is abandoned, although a residual inventory may be carried out later to determine if and when re-logging may be possible. Imposition of some harvesting regulations is not really forest management and is only workable in areas where human population density is very low and remaining areas of forest very large. Furthermore, repeated uncontrolled extraction of species as they become valuable on the international market can lead to structural and genetic degradation of the forests, which are then difficult to restore (Roche & Dourojeanni 1984).

Most forests are under more formal management systems, of which four main types may be recognized (after Gómez-Pompa & Burley 1991). Examples of management systems in these categories are listed in Table 2.1.

*Replacement systems*

This form of silviculture involves clear cutting and establishment of plantations, either of hardwoods or of introduced conifers or *Eucalyptus*, or of shifting agriculture.

Table 2.1. *Some well-researched forest management systems*

| Type and designation | Geographic location of example | Reference |
|---|---|---|
| **Replacement systems** | | |
| Plantation systems | Nigeria | Kio & Ekwebalan 1987 |
| | South-East Asia | Davidson 1985 |
| Taungya | Java | Wiersum 1982 |
| Mayan agroforestry system | Mexico | Gómez-Pompa *et al.* 1987 |
| **Clearing systems** | | |
| Malayan uniform system | Peninsular Malaysia | Wyatt-Smith 1963 |
| | Sabah, Malaysia | Chai & Udarbe 1977 |
| Timber stand improvement system | Philippines | FAO 1989 |
| Tropical shelterwood system | Nigeria | Lowe 1978 |
| | Assam, India | Nair 1991 |
| Mengo system | Uganda | Earl 1968 |
| Strip shelterbelt system | Peru | Hartshorn 1990 |
| **Natural regeneration systems[a]** | | |
| Selection management system | Malaysia | Lee 1982; Salleh & Baharudin 1985 |
| Modified selection system | Ghana | Asabere 1987 |
| Celos silvicultural system | Suriname | de Graaf 1986 |
| Periodic block system | Trinidad | Clubbe & Jhilmit 1992 |
| Queensland selective logging system | Australia | Shepherd & Richter 1985 |
| **Restoration systems** | | |
| Assisted natural regeneration | Mexico | del Amo 1991 |
| Enrichment (line) planting | Uganda | Dawkins 1958 |
| | Nigeria | Kio & Ekwebalan 1987 |

[a]Some of the listed systems are not true natural regeneration systems but closely approach the ideal. For example, the selection management system of Malaysia selects management options according to pre-felling inventory data; options include clear felling of dense commercial stands and subsequent replanting as well as polycyclic management of less densely stocked timber stands. Several of the listed systems includes post-logging stand refinement as a management option. *Source*: Adapted from Gómez-Pompa & Burley (1991).

Replacement of natural forest with industrial plantations is usually viewed as a means of improving yields of certain wood products and simplifying management. It does not yet involve large areas in the tropics. However, in less populated parts of the tropics there is considerable potential for the establishment of plantations on unproductive agricultural land as an addition to the gazetted forest estate. While products from plantations generally do not replace natural forest timber, their establishment may have the effect of reducing deforestation through provision of fuelwood, local construction materials, etc. (Kanowski *et al.* 1992).

Shifting agriculture is widespread through most of the tropics. Forest is substituted by a short-lived agroecosystem which is later abandoned and the land allowed to develop a secondary succession. The rate of forest regeneration on abandoned land depends on the seed bank, including seed rain from surviving standing trees and from surrounding forest areas, and coppicing of stumps. Derivatives of the simple shifting agriculture system include taungya, where seedlings of commercial trees are planted in with annual crops, and agroforestry systems where only the lower storeys of the forest are cleared and crops such as cacao are planted under the resulting shade trees (Nair 1992).

### Clearing systems

Clearing systems involve extensive modification of the timber stand following harvesting of standing commercial trees. Non-commercial stems may be removed by cutting, girdling or arboricide treatment, with the ultimate aim of creating an even-aged stand with commercial species predominating. These systems require the presence of good seedling stocks and retention of nurse trees to provide seeds for future crops. Long-rotation times implicit in these systems (perhaps 70 years to regenerate a crop of trees) have lately meant their substitution with polycyclic logging systems in most tropical forest areas, although such systems are still considered an option in some forests (e.g. Uganda: Rukuba 1992).

### Natural regeneration systems

These 'selection' or 'polycyclic' systems attempt to minimize necessary interventions through selection of generally low numbers of commercial trees for harvesting and protection of advanced growth. It is assumed that regeneration will occur naturally, without requiring extensive manipulation. The systems assume that the sizes and density of gaps formed during logging do not alter natural regeneration patterns and that sufficient seedlings of commercial species will become established to provide a second crop within the 20–30-year cycle commonly envisaged. The economics of these systems are problematic since capital return on the first harvesting operation is lower than with monocyclic systems. On the other hand, costs of long-term management are less.

Minimal intervention systems provide the best opportunities for incorporation of conservation-oriented management criteria. However, as pointed out by Whitmore (1990), although the theory is good and the practice promising, there is actually no conclusive evidence of their long-term sustainability.

### Restoration systems

In these systems management is introduced to regenerate productive forest on degraded land which may otherwise be locked into an arrested succession, as with the *Imperata* grasslands of South-East Asia. Forest severely damaged by poorly controlled logging operations may also require major intervention tantamount to restoration.

### Summary

Tropical forest products have been used throughout recent history by peoples living within or around the forest peripheries. The majority of such uses were originally non-destructive, although due to increasing human populations this may no longer be the case. Large scale and often destructive use of the forests is a comparatively recent phenomenon. Intensive timber logging began only with the introduction of power saws and hauling machinery in the 1940s. While non-timber products continue to be important in some tropical forest regions, tropical forestry is now largely synonymous with timber production.

Tropical forest management began in the early nineteenth century in India and Myanmar and rapidly spread into parts of Africa. Originally, management was implemented to protect strategic timber resources. More recently management has diversified into a variety of forms with various ends. These vary from replacement of natural forest with industrial plantations or agroforestry schemes (replacement systems), to extensive modification of the natural forest to increase productivity (clearing systems) to minimal disturbance and the use of natural regeneration to produce future timber crops (natural regeneration systems). Management may also be employed to regenerate forest on degraded land (restoration systems).

# 3

## Changes in the physical environment

### Introduction

Logging operations inevitably cause disturbance to the soil surface and to the remaining vegetation, which in turn affects hydrological cycles and can lead to erosion and sedimentation of water systems. This chapter considers how logging operations change the physical environment: the following chapter will consider actual changes to the forest vegetation.

The consequences of logging for the physical environment were first discussed in detail in a widely circulated but unpublished report by Ewel & Conde (1976). This drew attention to the many gaps in knowledge that existed. Amongst other topics, the report stressed the importance of basic research in:

(a) Evaluation of the effects of forest management practices on soil physical, chemical and biological properties, including impacts on soil microorganisms involved in nitrogen cycling, soil structure and tree nutrition (i.e. mycorrhizal fungi).

(b) The relationship between forest management practices and water quality, including chemistry, turbidity and temperature.

(c) The relationship between forest management practices and water yield, including impacts on the total amount, seasonal distribution and rate of return to the initial condition.

(d) Total-system nutrient budgets, including estimates of weathering and rainfall input rates, as well as leaching, runoff and biomass removal loss rates.

Since the 1970s, a considerable amount of research has been focussed on these topics, which together encompass most of the primary physical effects of logging except for the relationship between forest management practices and forest microclimate. However, the majority of research has been conducted in main timber source areas, particularly the dipterocarp forests of South-East Asia. The consequences of logging in more marginal forests, such as mangroves and seasonally flooding forests of the Amazon basin, have been very little studied.

### Microclimate

#### Types of microclimatic change

The lower levels of undisturbed forest are typically dark, humid, cool and wind-free. Sunlight penetrates into the forest where there are openings in the canopy, along water courses and where old trees have fallen. In these places the microclimate becomes hotter, drier and often more windy. In recently logged forest the proportion of early successional patches is greatly increased; much of the lower story is well-lighted, relatively dry, hot at midday, and experiences greatly increased wind turbulence. In Queensland forest where 22% of the canopy was lost during logging[0:16–23], the amount of light reaching the forest floor increased throughout the understorey, not just in the gaps themselves (Crome *et al.* 1992).

Perhaps the most important constraint on plant growth in natural tropical forest is the low level of insolation in the understorey. In tropical forest, 2% or less of the visible light striking the canopy reaches the forest floor (Chazdon & Fetcher 1984; Lee 1989). At least three-quarters of this consists of ephemeral sunflecks. The creation of gaps increases the amount of light striking the forest floor considerably, the increase being proportional to the size of the gap. In gaps of $< 200\,\mathrm{m^2}$ caused by a single treefall, light levels increased by five times (Brown 1993). In large gaps cleared for agriculture within the forest, light levels increased by 84 times (Schultz 1960). Changes in light levels persist until regrowth of plants fills the gap. In small gaps this may occur through sideways expansion of the canopy. This may reduce light levels more rapidly than in large gaps (after 2 years in a 400-$\mathrm{m^2}$ gap: Fetcher *et al.* 1984), even though seedling growth can be more rapid in the latter.

Increased insolation is linked to increased air and soil temperatures within gaps. Daily minimum air temperatures remain similar between gaps and adjacent forest but daily maxima increase markedly, generally by at least 4 °C (Fig. 3.1). The vertical stratification of temperature evident in undisturbed forest is largely lost in logged forests with high densities of gaps. Differences in soil temperatures can be even more marked. Soil at a depth of 2 cm experienced a diurnal temperature variation of 1 °C within natural forest, 5 °C in a small treefall gap and 14 °C in a clearing (Schultz 1960).

Within natural tropical forests relative humidity rarely drops below 80%, although much lower levels have been reported. Drops to 40% have been recorded recently in low-stature forests of the Tapajós Forest Reserve in central Amazonia (N. Brown, pers. comm.). In forests where $\geq 50\%$ of trees are destroyed by logging humidity levels of $< 50\%$ are common. The evaporative demand of the air was around seven times higher in a large gap than in the natural forest understorey in an example from Ghana (Longman & Jenik 1974).

The leaf litter and the soil to a depth of about 70 cm is mostly affected by microclimatic factors. (Below this depth the soil is mostly affected by hydrological factors: Bruijnzeel 1992.) The establishment of early secondary vegetation quickly reduces the rate of evaporation from the leaf litter and upper layers of the soil. However, the establishment of secondary vegetation dramatically increases transpiration losses, which may be significantly greater than the former evaporative losses. Overall, the moisture content of the air increases significantly once a vegetation layer is established and the extreme fluctuations are ameliorated.

## Cloud forest
This unique forest type of the Andes and Central America obtains around half its required moisture by 'occult precipitation', trapping moisture-laden air from the Pacific trade winds. The forest microclimate shows extreme seasonal fluctuations in humidity. Associated fluctuations in the moisture content of

Fig. 3.1. Variation in daily maximum and minimum temperatures across a recently formed gap in unlogged forest, Ulu Segama Forest Reserve, Sabah. Distances are measured from the base of a strangling fig tree whose crown was removed during a storm. (Source: Heydon 1994.)

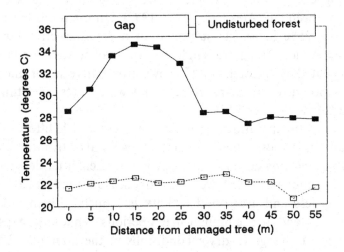

canopy organic material may be important in determining the distribution and reproduction patterns of epiphytes and associated canopy organisms (Bohlman *et al.* 1995). Tree felling in these forests can cause changes in the vertical structure of the vegetation, which affects microclimate through decreasing water filtration rates (Zadroga 1981). The biodiversity of cloud forests is linked to both vegetational structural diversity and the moisture gradient (Ralph 1985), either of which may be affected by forestry activities.

### Some direct effects of microclimatic change

Penetration of sunlight into lower levels of the canopy can adversely affect phytophobic plants, such as some epiphytes and mosses. These may only live in damp environments or may require high moisture levels to reproduce. Some forest trees are susceptible to bark scorch if exposed to direct sunlight. While seldom killing the tree this can encourage fungal invasion of the wood (Tinal & Palenewen 1978). Seedlings of many canopy trees are vulnerable to increased temperature and its associated water stress (Brown & Whitmore 1992).

Many trees in tropical forest have mycorrhizal associations (Brundrett 1991). The beneficial effects of mycorrhizal symbiosis have been poorly studied. Mycorrhizal association has certainly been reported to enhance uptake of certain nutrients (Janos 1980). However, the nutrients concerned are generally not limiting in tropical forest and the energy costs to the host in supporting the mycorrhizae need to be considered (Proctor 1992). It has been suggested (Smits 1985) that increased soil temperatures in large gaps may approach those lethal to mycorrhizae. Quantities of mycorrhizal spores in the soil were reduced by 75% in Malaysian logged[2:>70] forest where all trees ≥40 cm diameter at breast height (dbh) had been cut (Alexander *et al.* 1992). On the other hand, quantities increased by 25% in forests subjected to normal selective logging[?:50]. Die-off of some successional trees in large gaps may be attributed to their failing to be inoculated by mycorrhizae. However, the relative importance of mycorrhizae, as opposed to problems of heat and water stress affecting seedlings directly, remains undetermined.

Increased fluctuations in temperature and water content of the leaf litter and upper layers of the soil can seriously affect numbers and vertical distribution of litter invertebrates, thus affecting decomposition processes. Leaf litter decomposition rates fluctuate annually, being lower in the dry season when moisture content of the litter is less and arthropod activity is reduced (Levings & Windsor 1983). Litter thus tends to accumulate in dry periods. While large gaps remain in logged forest, a change to dryer conditions in the litter may be expected and thus an altered rate of litter decomposition.

Among vertebrate species, decreased humidity may limit the movements of amphibians to some extent, especially those that normally live away from water (e.g. the frogs *Microhyla* and *Rhacophora* in South-East Asia, and *Dendrobates* in the Neotropics). In temperate forests of the Appalachians, physiologically intolerant terrestrial salamanders of the Plethodontidae were either absent from recently clear-cut forest or forced underground until the canopy and leaf litter layer were re-established (Petranka *et al.* 1994). Salamander species-richness was highest in forests more than 120 years old (i.e. forests not subjected to any form of logging or other disturbance).

Heat stress can affect many species, although this is more likely to change foraging patterns than to limit their distribution in logged forest. Many tropical forest birds, particularly small insectivores, alter vertical patterns of foraging through the day. These birds may experience problems of water retention as upper layers of the canopy become dryer and hotter at midday (Bell 1982). Equally, however, birds may be following the vertical movements of their insect prey. Larger mainly insectivorous species of birds of paradise *Paradisaea raggiana* and *Cicinnurus regiius* also show vertical movements, while most frugivorous and mainly frugivorous species sit immobile in the canopy during the hottest part of the day.

Many understorey birds avoid crossing sunlit patches to avoid possible heat stress (Wong 1982; Karr & Freemark 1983). In Amazonian várzea, the understorey bird fauna of forested islands is depauperate, probably because few species are able to cross stretches of open water to recolonize after the annual floods (Remsen & Parker 1984). Furthermore, many understorey birds have large, bulbous eyes, adapted for foraging in dim light (Orians 1969). A large increase in light intensity may hamper their foraging efficiency or permit invasion of their foraging space by secondary forest species lacking this specialization.

Nocturnal species may experience heat stress while roosting. Very small bats, such as the blossom bats *Syconycteris* of New Guinea, have minimal (virtually no) temperature regulation capabilities. Species roosting in caves or treeholes are probably better protected against increased ambient temperatures than species such as the Neotropical banana bats *Thyroptera*, which roost in rolled immature *Heliconia* leaves. Species roosting in vegetation, under logs or under rock overhangs may be limited by a shortage of shaded roosting sites in logged forest, although this remains to be determined.

### Soil, water and nutrients
#### Soil damage and loss

Effects of road building and harvesting operations on tropical forest soils have been well studied (e.g. Soong *et al.* 1980; Borhan *et al.* 1987; Hendrison 1990). Between 12 and 30% of the soil surface is usually exposed as road surfaces, skidding tracks and landing areas. Damage to the soil is severe in all types of logging operation, both conventional (using skidders and tractors) and in overhead cable systems. Damage under the latter system is more localized, being centred on ridgetops where the hauling machinery is sited, but total soil areas damaged can be higher than using tractors (22.5% versus 18% in an experimental set-up in Malaysia: Borhan *et al.* 1987).

The use of heavy equipment of any kind results in some level of soil compaction, damage to the root mat and a decrease in infiltration capacity, all of which increase the possibility of erosion. The infiltration rate along newly vacated skid roads was found to be only 0.63 cm³/minute compared to 4.6 cm³/minute in the unlogged forest (Abdulhadi *et al.* 1981). Differences in infiltration rates can persist for many years (Fig. 3.2). The growth of seedlings on scoured and compacted soils is generally poor, although this appears to be due mainly to the loss of nutrients consequent to loss of topsoil rather than to soil compaction (Nussbaum *et al.* 1995).

The loss of soil from areas where forest is clear-felled can be on an enormous scale. The Ganges River has deposited enough silt into the Bay of Bengal to create the base of a potential extra 50 000 km² of land, largely due to deforestation of the lower slopes of the Himalayas. Soil loss rates are frequently reported to increase by > 25 000% following forest clearance.

Forest harvesting, where the soil is only partly exposed, is less damaging. In

Fig. 3.2. Average infiltration rates for topsoil (1) in undisturbed lowland rain forest, (2) in regenerating forest logged 12 years previously, (3) on a skidroad abandoned 12 years previously, Ulu Segama Forest Reserve, Sabah. (Source: Bruijnzeel 1992.)

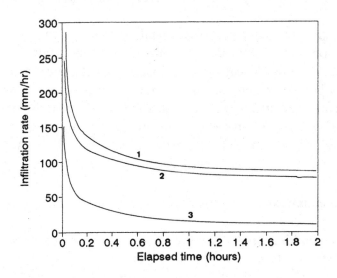

Sabah, logging[3:60] on a 35° slope resulted in the loss of an average of 45 mm of topsoil, equivalent to 454 m³/ha. Most loss occurs relatively soon after harvesting, however. Re-establishment of ground vegetation quickly reduces soil loss although levels may remain somewhat higher than in unlogged forest (Table 3.1). Extensive rill erosion leading to gully erosion generally occurs only on poorly sited and drained skidroads and feeder roads where erodable mineral soil has been exposed by bulldozing. On heavily damaged steep slopes, such as around overhead cable landings, landslips commonly occur as the decay of tree roots causes a reduction in soil shear strength (Hamilton 1985). Both gully erosion and landslips can be prevented by appropriate planning, or countered by remedial measures, but these are not always applied (Bruijnzeel & Critchley 1994).

### Water flow and sediment yield

Following removal of forest, atmospheric humidity, transpiration and particularly soil infiltration rates are much reduced. This results in an increased water yield, which can increase maximum flood levels while reducing water flow in times of drought. In the most extreme case, the annual flood crest of the Amazon mainstream at Iquitos, Peru, was never more than 26 m before 1970, but has never been less than 26 m since.

Changes in patterns of streamflow, using comparisons of paired catchments, are discussed by Bruijnzeel (1992). The amount of rainfall that runs off as saturation overland flow can increase 10 times following logging. This should increase peakflow and stormflow volumes considerably, although in practice these changes are not often significant. There may be a methodological

Table 3.1. *Effects of vegetation re-establishment on water run-off ('saturation overland flow') and rates of soil loss at South Lakimantou, Indonesia*

| Stage of logging | Water run-off (m³/ha per month) | Soil loss (tonnes/ha per month) |
|---|---|---|
| Undisturbed forest understory | 2 | 0 |
| Skidroads newly constructed but unused | 149 | 10.8 |
| Skidroads in operation | 189 | 12.9 |
| Two years after end of harvesting | 43 | 6.2 |
| Three years after end of harvesting | 19 | 3.2 |

Tree loss rate resulting from logging was c. 49%. Plots located on slopes of 8–10°; annual precipitation 2429 mm.
*Source*: Adapted from Hamilton (1985).

problem in the failure of comparative studies to record significant changes in flow volume. Logging contractors may well adjust normal logging procedures where hydrological studies are known to be underway (Douglas *et al.* 1990).

Sediment yields increase by between two and 10 times following road construction and may be 20 times as high following harvesting. In one study in Sabah sediment yield was, however, down to 3.6 times the unlogged forest level only 12 months after the completion of logging[1:57] (Douglas *et al.* 1992). The rate of decline is typically steep over the first 2 years as ground vegetation becomes established, but tends to level off at a higher value than in unlogged forest (Bruijnzeel 1992). Sediment washed off by the first storms following the onset of harvesting activity accumulates in stream beds and may then take some time to be fully excavated, perhaps as long as 5 years. Rill and gully erosion of exposed mineral soil may provide an even longer-lasting supply of sediment.

### Nutrients

Nutrient concentration in the topsoil within small treefall gaps (less than about 150 m$^2$) has not been found to differ from that in the topsoil of surrounding mature forest. In larger gaps, results are varied. On Maracá Island, Brazil, artificial gaps as large as 2500 m$^2$ had no reported effect on nitrogen mineralization (Marrs *et al.* 1991). In gaps of 500–2500 m$^2$ in Costa Rica, concentrations of nutrients, including nitrates, in soil moisture showed a definite increase attributable to the release of nutrients from decomposing tree crowns and other material (G. G. Parker, in Bruijnzeel 1992). This increased nutrient level persisted only for 12 months, after which time the nutrients were absorbed by pioneer growth. There may thus be natural fluctuation in nutrient contents associated with treefall events.

Logging not only increases the density (and average size) of treefall gaps, but influences the cycling of nutrients by removing some from the system. Nutrients may be removed in two ways. First, in the form of organic material: harvested logs or partially decomposed litter which can be washed away by increased overland water flow. Second, through enhanced leaching of soluble nutrients by increased water runoff at a time when nutrient release, at least in large gaps, may be high.

In an intensive but well-managed operation in Queensland, quantities of most nutrients in the soil did not alter significantly during the first 4 years following logging[1-4:30] (Gillman *et al.* 1985). The exceptions were calcium, magnesium and potassium, which reflected an overall 15% drop in the level of organic carbon. This was attributed to the removal of stemwood as logs. On the basis of nutrient input from bulk precipitation, it was calculated that it would take up to 34 years for the various nutrients to regain pre-logging levels (calcium taking the longest). Since a 40–50 year cycle for removing logs was envisaged, the removal of these nutrients was considered sustainable. A similar study in peninsular Malaysia also found recovery of calcium levels in the soil to be problematic, and suggested phosphorus levels should also be monitored (Amir *et al.* 1990).

Two studies, in Suriname and peninsular Malaysia, have used calculations from changes in water chemistry (rather than changes in the soil) to calculate nutrient budgets over proposed logging cycles (Poels 1987; Y. Zukilfi, in Bruijnzeel 1992). In the Suriname example, a net loss of calcium is predicted over the scheduled 20-year rotation. In the peninsular Malaysian example, the 30-year cycle will result in net losses of both calcium and magnesium (Table 3.2). The question that remains unresolved is whether these loss rates will necessarily limit the rate of forest regrowth. Trees may well contain more nutrients than they actually require (Proctor 1992). Results from the Suriname and peninsular Malaysian sites concern removal of stemwood only at levels that are low and moderate by international standards (23 and $50\,\text{m}^3/\text{ha}$). Removal of whole trees, rather than just stemwood, can result in a substantial increase in nutrient loss (Mann *et al.* 1988). In monocultural tropical

Table 3.2. *Approximate nutrient budgets for forestry operations in Suriname and peninsular Malaysia*

| | Nutrient concentrations (kg/ha) | | | | | | |
|---|---|---|---|---|---|---|---|
| | Suriname | | | Peninsular Malaysia | | | |
| | K | Ca | Mg | K | Ca | Mg | Total N |
| Input into the system | | | | | | | |
| Precipitation (per year) | 2.5–4.5 | 2.5–3.0 | 1.3–2.8 | 5.7 | 3.8 | 0.7 | 11.3 |
| Export from the system | | | | | | | |
| Harvested logs (total) | 35 | 80 | 7 | 45 | 200 | 20 | 70 |
| Extra leaching (total) | 12 | 30 | 11 | 75 | 30 | 15 | ? |
| Years required for nutrient levels to be regained | 10.5–18 | 37–44 | 6.5–14 | 21 | 60 | 50 | >6.2 |

Tree loss rates resulting from logging were 17% in Suriname and 50% in peninsular Malaysia. Data assumes total reabsorption of nutrients from logging debris (leaves, branchwood, etc.).
*Source*: Adapted from Bruijnzeel (1992); original data from Poels (1987) and Y. Zukilfi.

plantations, where a high percentage of biomass is removed, nutrients can quickly become limiting. At Jarí in Brazil the need for addition of fertilizers between crops is well established.

### Some direct effects of soil damage and loss of water quality

Soils compacted by harvesting machinery, areas where soils are scoured by construction of roads and log landings, and the various less disturbed microsites in logged forest gaps may all develop different plant regeneration successions. Heavily disturbed sites may first be colonized by grasses, bamboos or stemless palms (such as *Eugeissona* in Malaysia). In peninsular Malaysian dipterocarp forests subjected to high-lead logging[1-14:51], for example, large areas of heavily damaged soils are colonized first by the fern *Gleichenia* and hooked creeping plants such as *Leuconotis* which encroach into the damaged region from the edges. Smaller areas of less damaged soils tend to be colonized by bananas and gingers, interspersed with a variety of pioneer and climax tree seedlings. The level of soil damage can be a principal determinant of the speed at which tree cover is regained.

Disturbance of the soil through scouring or compaction will obviously affect the soil fauna to a great extent. Besides the microorganisms that live permanently in soil, a number of larger animals rely on this substratum (e.g. burrowing snakes *Typhlops*). At the Tekam Forest Reserve, peninsular Malaysia, burrow systems and underground refuges used by a variety of animals, especially snakes, were extensively damaged during logging[0-1:51] (Johns 1983). Encounter rates with the usually cryptic forest snakes are much higher in areas undergoing logging. Mortality is also high, due both to crushing by logging machinery and to an increase in the numbers of predators (particularly large raptors). Most logging contractors consider this no bad thing.

Increased sediment loads in watercourses directly limit animal species which require clear, oxygen-rich water either as a permanent habitat or for breeding. High sediment loads are temporary, but there may be a critical period immediately following logging when aquatic animals must migrate to clear-water refuge areas, perhaps to main rivers. The effects of logging operations on freshwater ecology have been little studied. Samat (1993) reported that siltation of watercourses in Sabahan logged[1-2:57] forest caused severe declines in diversity and abundance of bottom-feeding fish, whereas open-water fish were less affected.

At the Tekam Forest Reserve, the creation of stagnant pools where water courses are blocked, and the creation of small ponds in depressions caused by logging machinery, cause a distinct shift within the amphibian community

(Johns 1983). Frogs typical of clear, swift-flowing forest streams (*Rana hoseii, Ansonia* sp., *Megophrys monticola*) were infrequently observed following logging[0–1:51]. These species were replaced by a guild of secondary forest ranids (*Rana erythraea, R. chalconota, R. limnocharis, R. nicobariensis*) and a rhacophorid (*Polypedates leucomystax*) all of which breed successfully in largely anoxic water. Mycrohylids such as *Microhyla heymonsi*, which breed in small ephemeral pools on the undisturbed forest floor or in water-filled elephant footprints, were able to persist in logged forest using small pools formed after rainstorms. Such pools also form breeding sites for mosquitos (Culicidae) and an increased abundance of these insects is a common consequence of logging operations.

Increasing silt loads in major rivers, can result in offshore silt dumping. This has killed coral reefs in several parts of the Indo-Pacific region. Choking of water flow through mangroves by excessive silt dumping disrupts the lifecycles of many marine animals, such as prawns and fish which spawn in this habitat, and can thus also affect offshore fisheries (Whitten *et al.* 1984).

### The special case of periodically inundated forests

Sedimentation of medium or fine particles (mud rather than sand) along coasts or major river systems can support specialized forest types. In coastal regions throughout the tropics, building mudflats may be colonized by sediment-trapping plants such as *Avicennia* and *Rhizophora*. Through succession this may develop into mangrove forest. Along the Amazon-Solimões river drainage in Brazil, seasonal changes in water level cause extensive flooding by silt-rich water originating in the Andean mountain range, and a seasonally flooding *várzea* forest has resulted. A dependence on the dynamics of water flow and sedimentation can mean that these forest types are less resilient to environmental effects of forestry operations than are most dry land tropical forests.

#### Mangrove forest

Mangrove forests have a low floristic diversity. Only 90 species of woody plants have been recorded of which 55 are restricted to this vegetation type. The most diverse and extensive mangroves are in the Indo-Pacific region.

Being coastal, mangroves are easily accessible and they have long been commercially important. The Matang mangrove forest in peninsular Malaysia and others in the Sundarbans (Bangladesh and north-east India) have been managed for timber production since the beginning of the twentieth century. Mangroves are most often harvested to provide charcoal (*Rhizophora* is well suited to this end use) and poles (used in piling and scaffolding as the wood is very hard and termite resistant). Some woodchip operations are underway in the Indo-Pacific, and their viability is being studied in the upper Amazon region. Few mangrove trees are suitable for commercial timber, only *Hereteria fomes* of the Indian Ocean region being marketed extensively. The sustainability of timber offtake from managed mangrove has been called into question. At Matang, mangrove forest yielded 299 tonnes/ha on its first felling, but this has declined to 150 tonnes/ha from managed forest during the 1960s and 136 tonnes/ha during the 1970s. Enrichment planting is now required to maintain yields at an economic level (Tang *et al.* 1981).

Most logging operations in mangrove are intensive and the end result close to clear felling (despite minimum girth limits set for harvestable trees). The mud-trapping function of the stilt rooted trees is disrupted, surface mud tends to be eroded and higher elevated soils tend to dry out. The drying soils become extremely acidic (C.P. Burnham, in Whitmore 1984) which retards the regeneration of mangrove trees. Drying soils tend to be colonized by *Acrostichum* ferns, which are uncommon in primary mangrove forest.

In Sarawak, severely damaged mangrove may be invaded by mud-lobsters (*Thalassina anomala*). Mud-lobsters dig volcano-like mounds which may further disrupt water flow, elevate soils to cause further erosion and drying, and further promote the establishment of *Acrostichum* over mangrove tree seedlings. As long ago as the mid-1970s it was pointed out that management of these mangroves for continued timber and pole production was already synonymous with replanting to keep ahead of the growth of *Acrostichum* (Christensen 1978). Mangrove tree seedlings had to be sufficiently established to out-compete the *Acrostichum* before it covered the ground between mud-lobster mounds and choked the system of water flow.

### Várzea

The eastern slopes of the Andes receive in excess of 2500 mm of rain annually, with marked seasonality. This causes water level fluctuations of up to 15 m in the Amazon-Solimões system of western Brazil, flooding river valleys up to 100 km wide. The silt-laden white-water rivers flood around 70 000 km$^2$ of land, which has developed an association of vegetation types collectively termed várzea.

The dynamics of várzea formation are discussed by Ayres (1986b). Within várzea, tall forest is restricted to alluvial levées, marking the deposition areas of old watercourses. The forest is flooded for less than 6 months each year. Separating the interlinked forest corridors ('restingas') are areas of low dense scrub or grass ('chavascais'). These areas are flooded for more than 6 months each year and often submerge completely. During peak flood levels, only the crowns of the restinga trees appear above the water.

Várzea forest is typically less species-rich than adjacent dryland forest. It contains few, if any, unique tree species, but the relative abundance of species is quite different to that of dryland forest. Many várzea forest trees are harvested to supply domestic timber markets as they are easily accessible. Logging has occurred throughout most of the Amazonian várzea for around 30 years, and is becoming progressively more intensive as the number of marketable tree species increases (Ayres & Johns 1987). Although damage levels associated with logging in várzea are quite low, the restinga forest has very specific hydrodynamics which are easily affected by tree felling.

Once established on the banks of watercourses, várzea trees grow impressive spreading buttresses, which slow the passage of floodwater and cause it to drop suspended sediments, including nutrients valuable to the tree. The longer the trees are established, the more alluvia is trapped and the higher the restinga rises. Ribbon lakes and channels constantly change course, however, as treefalls breach restingas or old watercourses silt up. The felling of many trees during logging accelerates the rate of change of watercourses and alters patterns of flow and silt deposition. Even single treefall incidents may divert channels and cause erosion of levées, or prevent adequate drainage of low-lying forested restinga and cause additional tree mortality through permanent flooding. In the long term, the balance of restinga and chavascal may be changed, with chavascal predominating.

### Some wider implications of environmental change
#### Climatic shifts
Since the 1970s, removal of forest cover over large land areas has been predicted to cause apocalyptic changes in local or even global climatic conditions (Potter 1975). The potential effects of widespread change from forest cover to grassland or annual crops have since been extensively modelled (IPCC 1990, 1992). Models have been refined through ground studies in Amazonia which have helped quantify both the atmospheric and surface-related processes that can give rise to local climatic shifts (Shuttleworth & Nobre 1992).

Burning or re-burning of forest land to create and maintain pasture, as occurs particularly in Amazonia and central America, can cause significant loading of the atmosphere with smoke aerosols. Deforestation also leads to substantial releases of carbon, in the form of carbon dioxide, into the atmosphere.

Smoke aerosols can significantly alter the absorption properties of the atmosphere (Dickinson & Kennedy 1992). However, since chemical properties of smoke vary considerably, it is difficult to predict effects on local climate. These aerosols tend to be removed from the atmosphere over only a few days through dry deposition (gravitational settling) or by rain. Their direct effects on climate are broadly co-located with their source regions and proportional to rates of burning in those regions. Their effects are short-lived once burning stops.

Carbon dioxide, on the other hand, is of more significance to global climatic change in view of its much longer atmospheric lifetime (50–200 years) and its importance as a 'greenhouse gas'. Emissions of carbon from tropical deforestation throughout the 1980s are estimated to have averaged $0.6$–$2.6 \times 10^9$ tonnes/year compared with a contribution from fossil fuel burning of $5.5 \times 10^9$ tonnes/year over the same period (IPCC 1994). Global climatic models (e.g. IPCC 1990) suggest that a rapid global warming of several °C will occur in the next century as the concentrations of such greenhouse gases build up. This in turn might seriously threaten sensitive ecosystems worldwide (IPCC 1990), including tropical forests. Sequestration of atmospheric carbon begins once a vegetation cover is re-established on deforested land, but is enhanced if the vegetation is a long-lived carbon store, such as a new forest. This has been put forward as an argument for reforestation of large areas of degraded land in the tropics (Sedjo 1989), the impetus for which may well be provided by 'energy taxes' imposed on fossil-fuel burning industries, particularly energy companies in developed countries (Marsh 1993).

Large scale deforestation may also have substantial impacts on the climate of a deforested region through its direct effects on the local energy and moisture balance, and resulting feedbacks. Higher surface albedo resulting from conversion of forest to grassland would initially reduce the solar energy absorbed at ground level, but reductions in cloud cover may subsequently counteract this (Henderson-Sellers *et al.* 1993; Lean & Rowntree 1993). Substantial changes in the hydrological cycle are predicted in deforestation studies with global climatic models. Reduced precipitation and evapotranspiration rates in direct response to the changes are likely. Atmospheric circulation patterns could be affected over a wider area than that which is actually deforested (Henderson-Sellers *et al.* 1993, Lean & Rowntree 1993). Local climates may shift towards longer dry seasons (Nobre *et al.* 1991). Rain forests in particular may regenerate successfully only in the absence of pronounced dry seasons.

Unlike deforested land, logged forest may well maintain surface energy and moisture exchanges. This is because it maintains a dense, if uneven, canopy and retains at least some deep-rooted trees. The latter is particularly important in maintaining transpiration rates during irregular dry periods, helping to regulate the hydrological cycle and cloud cover.

### Susceptibility to burning

The extent to which surface energy exchanges are maintained in logged forest depends on the degree of canopy loss. The increased susceptibility of heavily logged forest to drying and subsequent fire outbreaks is well established (Uhl & Buschbacher 1985). Even where considerable numbers of deep-rooted trees remain, soil water can be exhausted during long periods of drought and drying can eventually occur. The aberrant El Niño/Southern Oscillation episode in the Pacific region during 1982–83 resulted in unprecedented droughts in several parts of the world (Gill & Rasmusson 1983), and extensive forest fires in logged areas of East Kalimantan and Sabah (Beaman *et al.* 1985; Leighton & Wirawan 1986). Recently logged forest was markedly more susceptible to fire outbreaks than old logged or primary forest. In Sabah, burning destroyed 5.6 times as much logged forest as primary forest. In East Kalimantan, twice as much logged as primary forest was destroyed. The latter ratio was lower due to part of the primary forest being killed by the extended drought (not by burning). Post-fire leaching and erosion can add significantly to the loss of nutrients from regenerating forest.

### Summary

Forestry operations induce a variety of changes in the physical environment, affecting microclimate, soils, water, and nutrients. The regenerating forest tends to be dryer, hotter at midday and to show increased light levels. Soils may be compacted, decreasing infiltration rates and leading to erosion of exposed areas. Increased water runoff can result in cycles of flooding and drought, with sedimentation of watercourses through increased silt dumping. Nutrient levels in the soil and water may be reduced through removal of biomass from the system, although providing total biomass removal is low this is not likely to be limiting to plant regrowth.

Physical changes resulting from logging are potentially most damaging in periodically inundated forests which are dependent on the dynamics of water flow and sedimentation. Certain types of logging disturbance may favour the establishment of secondary growth rather than regeneration of forest.

Many species of plant and animals are directly affected by changes in the physical environment. Lower plants, mycorrhizae and seedlings of forest trees may be vulnerable to increased insolation, increased temperatures and associated water stress. Microclimate changes may also affect animals requiring high levels of humidity, such as amphibians. Soil compaction and erosion seriously affects the soil microfauna, and can be damaging to species requiring burrows or underground refuges. Sedimentation of river systems adversely affects many aquatic species.

The entire tropical forest ecosystem is potentially threatened by climatic change resulting from the loss of forest cover. This mostly affects heavily deforested regions, however. The retention of a large percentage of forest cover as logged forest maintains surface energy and moisture exchanges. However, microclimatic changes in logged forests may make them more susceptible to forest fires.

# 4

## Forest regeneration and gap dynamics

### Introduction

The impact of logging operations on tropical forest vegetation is reliant on two main factors, the number of trees removed and the care taken in doing so. The extent and type of any interventions applied subsequent to felling also affect the regenerative capability of the forest.

While tropical forests contain a very large number of tree species, relatively few are acceptable to the timber trade. Tropical forest timber logging is inevitable 'selective' to some degree. In peninsular Malaysia, for example, there are around 2500 tree species of which 700 reach a utilizable diameter. Of these 400 are considered to have commercial properties, but only about 30 are exported in significant quantities. The degree of selection is greater than this in many tropical countries. In parts of central Africa only a single species, *Aucoumea klaineana*, may be cut for commercial use.

However, felling timber trees is not the main cause of damage in tropical forestry operations. The majority of tree loss is caused by construction of access roads and particularly by the movement of hauling machinery between cut logs and loading points. This so-called 'incidental damage' inevitably greatly exceeds actual felling damage. Overall damage levels are much influenced by the extent to which hauling machine operators attempt to minimize damage during this stage of the logging operation (Crome *et al.* 1992). The degree of damage sustained during felling and log removal is critically important and determines the degree of change in forest structure and tree species composition, the size and frequency of gaps and the success of regeneration processes.

This chapter documents the extent to which the forest vegetation is affected or changed by logging. The importance of reducing damage levels sustained during logging and how this might be implemented will be discussed at a later stage.

### Damage levels

Felling intensities vary considerably within and between geographical areas. In Malaysian dipterocarp forests dominated by consociations of *Dryanobalanops aromatica* or *Shorea curtisii* the felling rate may reach 72 trees/ha, equivalent to a stump every 12 m. Similar levels can be reached in the *S. johorensis–Parashorea melaanonan* forests of eastern Sabah. In most South-East Asia forests, however, extraction levels are 14–24 trees/ha, giving a stump every 20–27 m. In most Amazonian *terra firme* forests extraction levels do not exceed 3–5 trees/ha. In central Africa < 1 tree/ha is currently cut in forest areas remote from main transport routes.

Felled trees are generally large emergents with crowns of up to 20 m in diameter. Their falling may cause considerable damage to other trees through a 'knock-on' effect. In Uganda, felling trees of > 5 m$^3$ log volume has been reported to damage areas of up to 2000 m$^2$ through this knock-down effect, equivalent to a circle of 50 m diameter. In Pará, Brazil, felling single mahogany trees with mean log volumes of 5 m$^3$ causes a mean loss of 31 other trees and felling gaps of 1100 m$^2$ (Veríssimo *et al.* 1995). In Suriname, the area covered by single felled trees varied from 100–600 m$^2$ and averaged 225 m$^2$ (W. H. H. Mellink, in de Graaf 1986).

Since many emergent trees have considerable heights of clean bole, damage will be minimal along most of the length of the fallen bole and most severe where the crown hits the ground. Felling 10 trees/ha will cover around 25% of the ground with fallen crowns. Much of the canopy gap will be directly above the fallen bole, however, and most of the seedlings and saplings in this area will remain undamaged. In a treefall gap studied in Sabah (Brown 1990) 73% of seedlings remained undamaged. Most seedling damage in logged forest occurs during extraction of the bole.

In most cases, natural treefall gaps are filled by growth of existing seedlings or colonizing trees. Occasionally, larger and longer-lasting gaps may be created by sequential treefalls, one falling tree pushing over several others or trees around the gap subsequently falling as a result of windthrow. In an extreme case, sequential windthrow has created narrow gaps up to 8 km long in Sarawak peat swamp forest (Whitmore 1990).

During harvesting operations, damage caused by felled trees is a relatively small proportion of overall damage levels. Initial road construction, prior to felling, often includes clearing of 12–20-m wide strips each side of the road to facilitate drying of the surface after rain. Log loading areas are also cleared, as are log landing areas if overhead cable systems are to be employed. Pre-felling clearance for the loading and transport network can destroy 6–20% of the forest. This varies largely according to the density of access roads, which itself reflects the volume of timber to be harvested.

In intensively logged[1:65] forest in peninsular Malaysia, skid roads, made by tractors moving between cut logs and main access roads, can average 4 m in width and may total 27 km in length per km² of forest. Tractors avoid large trees, and thus do not further affect the structure of the canopy, but destroy seedlings and saplings on 11% of the forest floor, in the above example. Studies in dipterocarp forest have usually shown that 30–40% of the forest floor may be left bare of vegetation as a result of logging[1:54-65] activities (e.g. Kamaruzaman 1991; Nussbaum 1995).

Loss and damage to residual trees is extremely high in most operating logging systems (Table 4.1). Loss and damage to seedlings can be equally high, both through physical influences (direct mortality through roading activities, etc.) and as a result of changes in microclimatic conditions. For example, studies in South-East Asian and West African forests have shown that 30–44% of the residual seedling stock may be killed or damaged as a result of being covered with logging[1:55-65] debris. In peninsular Malaysia, mortality of seedlings varied according to harvesting technique but there was no clear relationship between size class and mortality level (Table 4.2).

The loss of trees during felling and extraction is exacerbated by the subsequent death of many damaged trees. This occurs through fungal infection as a result of damage to the protective bark and cambium, and particularly through increased windthrow. The uneven nature of the canopy in logged forest increases wind turbulence, to which shallow-rooted mature trees are particularly susceptible. A single violent storm at Ponta da Castanha, Brazil, caused 19 treefalls/km² in logged[11:61] forest (with an average of 4.7 trees ≥30 cm girth destroyed per treefall) compared with 2.2 treefalls/km² in adjacent unlogged forest (Johns 1986a).

### Changes in forest structure

Harvesting operations result in an irregular pattern of clumps of almost undamaged forest, damaged forest (with large quantities of logging debris on the forest floor) and open gaps along roads and where trees have fallen (Fig. 4.1).

Although only the largest trees of particular species are cut for their timber, small timber trees and non-timber trees of all sizes are also destroyed and damaged. In a study in peninsular Malaysia, the frequency distribution of size classes of tree ≥30 cm girth was not significantly different following logging[1:51] (Johns 1988). The mortality of all tree size classes was around 50%.

Similar results were obtained in Queensland (Crome *et al.* 1992). As pointed out by Crome *et al.* (1992), however, consideration of very large sample sizes can indicate reductions in the proportion of large trees in the forest, especially where management interventions remove non-commercial large trees. This can

Table 4.1. *Causes of tree mortality during harvesting operations*

| | Loss of trees (%) | | | | | |
|---|---|---|---|---|---|---|
| | Gunung Tebu, Malaysia[a] | South Pagai, Indonesia[b] | Queensland[c] | Tekam Malaysia[d] | Nigeria[e] | Ponta da Castanha, Brazil[f] |
| **Killed** | | | | | | |
| Timber trees | 10 | 8 | 5 | 3.3 | 1 | 0.6 |
| Destroyed during construction of access roads and landing areas | 55 | 46 | 15 | 8.4 | 18 | 60 |
| Destroyed during felling and skidding | | | | 39.2 | | |
| **Remaining** | | | | | | |
| Damaged | 35 | | 20 | 6.0 | 7 | |
| Undamaged | | 46 | 60 | 43.1 | 74 | 39.4 |

Results from Queensland predate the introduction of strict silvicultural rules in 1982.
*Sources*: [a]Burgess (1971); [b]Whitten *et al.* (1984); [c]Whitmore (1990); [d]Johns (1988); [e]Redhead (1960); [f]Johns (1986a).

Table 4.2. *Mortality of seedlings and saplings due to conventional and high-lead logging operations: peninsular Malaysia*

| | Mortality (%)[a] | |
|---|---|---|
| Size class | Conventional logging (tractors) | High-lead logging |
| **Seedlings (cm height)** | | |
| <15 | 52.5 | 47.5 |
| 15–29 | 25.6 | 47.1 |
| 30–89 | 27.8 | 55.4 |
| 90–149 | 40.9 | 62.0 |
| 150–299 | 36.7 | 64.6 |
| 300–5000 | 32.0 | 58.5 |
| **Saplings (cm dbh)** | | |
| 5–10 | 26.4 | 58.9 |

[a]Mean of three experimental plots, minimum cutting limits 45–60 cm dbh. Loss rates of trees > 10 cm dbh was around 51% (study plot combining both logging methods).
*Source*: Adapted from Borhan *et al.* (1987).

particularly affect the abundance of 'snags'[1] of both commercial and non-commercial species.

Basal areas of logged forest are reduced proportionately to tree loss rates, which reflects the fact that tree loss is random over all size classes in intensive logging operations. In Malaysia and Indonesia, a 50–60% reduction in basal area under both polycyclic and monocyclic systems is usual. However, re-growth can be rapid. The mean tree height and trunk diameter were 2 m and 2 cm, respectively, in 1-year-old logged[1;55] dipterocarp forest in Sumatra but had increased to 22 m and 35 cm in 10-year-old logged[10;55] forest (Geollegue & Hue 1981). Following light semi-mechanized logging[8–9; <50] in dry deciduous forest in Madagascar, the mean size of canopy tree was reduced but

Fig. 4.1. The spatial distribution of logging damage in a section of compartment C13C, Tekam Forest Reserve, peninsular Malaysia. Damage is scored as the percentage of trees ≥ 30 cm girth at breast height (gbh) destroyed in each 50 × 50 m quadrat. Heavy black lines represent the position of logging roads, sited along ridgetops. (Source: Johns 1983.)

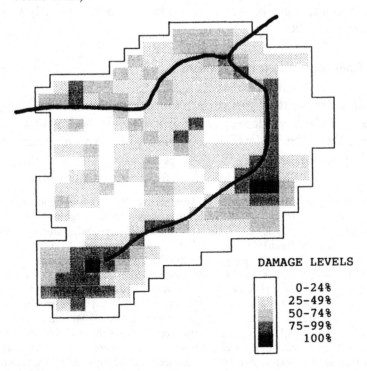

DAMAGE LEVELS

0–24%
25–49%
50–74%
75–99%
100%

---

[1] Following American terminology, the term 'snag' is used to denote a standing dead or dying tree. This is distinct from 'snagging' and 'snags' which may also be used to refer to cut trees that become entangled in the surrounding canopy and are thus prevented from falling to the ground.

changes in most vegetation structural variables could not be separated from natural variation due to forest patchiness (Ganzhorn *et al.* 1990). On the other hand, very old logged[40-60:>50] forest at Budongo, Uganda, remained structurally different to unlogged forest (Plumptre *et al.* 1994).

The average annual increment of standing timber in unlogged dipterocarp forest is 1–2 m³/ha per year (equivalent to loss rates, since the system is stable). Increment achieved in logged forest is very variable, despite the aim of increasing the increment of commercial timber species common to most silvicultural systems. In the Philippines, a mean increment of 3.2 m³/ha per year (range 1.6–8.6) was achieved from forests averaging 75 m³/ha timber production (Rapera 1978). In the Tapajós Forest of Brazil, where 75 m³/ha was removed in 1979, increment was 6 m³/ha per year 3 years post-logging but had decreased to 4 m³/ha per year 13 years post-logging, of which the commercial increment was only 1.8 m³/ha per year (Silva *et al.* 1995). Commercial timber volumes had reached 54 m³/ha after 13 years, suggesting successful regeneration, but this was actually largely due to the acceptance of many additional tree species on the market. At extremely high levels of timber production, and associated damage, negative growth rates may result in the early years after harvesting (Miller 1981).

### Changes in forest composition
*Community level changes*

If small areas of forest are compared, tree species richness and diversity tends to be less in logged forest. For example, both parameters show reductions where 50 m² plots are compared between unlogged, recovering[11:?<35] and degenerating[12:?<50] logged forests in the Kibale Forest, Uganda (Table 4.3).[2] The high rates of tree loss in intensive logging operations

[2] A number of studies of the effects of logging on biodiversity have been conducted at Kibale. These usually compare unlogged forest plots with what are termed 'lightly logged' (14 m³/ha timber removed) and 'heavily logged' (21 m³/ha timber removed) sites. Both stated logging intensities may actually be regarded as 'light' by international standards. In actual fact, however, timber volumes removed, particularly from the latter site, are considered to be underestimates as quite large amounts of cut timber may have been undeclared (i.e. stolen) and thus do not appear in the records (T. T. Struhsaker, pers. comm.). There is a clear difference between the former site, where basal area reduction 11 years post-logging was only 15% and the latter where basal area reduction 12 years post-logging was 60% (Skorupa & Kasenene 1984). While the former site appears to be recovering, the latter site is degenerating due to extension of canopy gaps through windthrow and frequent visitation by elephants which are retarding tree regeneration. The difference in regeneration success between these sites cannot be related to the stated logging intensities, which are probably incorrect, nor to differences in damage levels during logging, which are unrecorded. This illustrates a typical problem in *post facto* logging studies: causal relationships between forestry practice and ecosystem changes cannot be determined accurately since essential data are missing.

are statistically likely to eliminate rare tree species from small study plots, even if these are not commercial species. On a larger scale, however, there may not be a reduction in species richness or diversity. Results from Queensland suggest that logging reverses a natural loss of diversity that occurs as a forest matures after disturbance and pioneer species die out; certainly there is no significant difference between species diversity in unlogged and logged[1-17:1-34] forest plots (Fig. 4.2). In the Budongo Forest of Uganda the successional mahogany forest resulting from logging[1-60:>50] is considerably more diverse than the climax *Cynometra* forest (Plumptre *et al.* 1994).

### Changes in species composition

In cases where logging is highly selective for a rare species, and seed trees are not protected, that tree can become extirpated over large forest areas. This has happened with the highly valuable Brasilwood *Caesalpinia echinata* over large areas of Amazonia. Similarly, in the Tapajós National Park of Brazil, almost all large specimens of a single tree *Aniba duckei* have been removed. In such cases, the densities of one or a few target trees are reduced but the densities of other trees remain unchanged. Forest composition thus changes very little.

In mechanized logging operations where damage levels are significant, changes in species composition might be expected. In fact, results from peninsular Malaysia demonstrate that the proportional representation of different families and genera of trees does not change immediately following

Table 4.3. *Comparative measures of tree species richness and diversity in small unlogged and logged forest plots at Kibale Forest, Uganda*

| Species richness and diversity measures | Unlogged[a] | Regenerating logged forest | Degenerating logged forest |
|---|---|---|---|
| Total tree density (stems >9 m tall/ha) | 256 | 267 | 125 |
| Mean no. species recorded in 100 stem sample | 25.6 | 23.9 | 18.2 |
| Mean no. species enumerated in 5 × 50 m² plots | 25.3 | 22.7 | 14.3 |
| Equitability (Hill's evenness measure)[b] | 0.62 | 0.53 | 0.50 |
| Diversity (Shannon–Wiener index)[b] | 2.76 | 2.48 | 2.21 |

[a]'Unlogged forest' had actually been subjected to a low level of pit-sawing, but this was not thought likely to influence results.
[b]Based on results from species enumeration in 5 × 50 m² plots.
*Source*: Adapted from Skorupa (1986).

logging[1:51]. This is once again a result of the random nature of damage during felling and skidding operations (Johns 1988). Significant changes in tree species composition do occur in time due to the establishment of pioneers. A study in western Kalimantan, Indonesia, has also suggested differential mortality of residual tree species post-logging may give rise to changes in species composition even where pioneers are discounted (Cannon *et al.* 1994).

Pioneer trees germinate and grow extremely rapidly following logging, some species may reach a girth of 30 cm 5–6 years after logging. There are characteristically few species of pioneer in tropical forests relative to the numbers of climax species and they belong mostly to only a few families (particularly the Euphorbiaceae). The genus *Macaranga* (Euphorbiaceae) is particularly successful in South-East Asia, as is *Cecropia* (Urticaceae) in the Neotropics, *Musanga* (Urticaceae) in Africa, and the pantropical *Trema* (Ulmaceae). In intensively logged forest, pioneers can dominate the regenerating understorey. At Ulu Segama, Sabah, 25% of logged[12:57] forest area was dominated by pioneer vegetation. At the Tekam Forest Reserve in peninsular Malaysia, the percentage representation of the Euphorbiaceae in the tree community (stems >30 cm girth) rose from 27% in unlogged forest to 38% after logging[5–6:51]. Almost all of this increase was accounted for by *Macaranga*.

Changes in the abundance of other plants has not been well researched.

Fig. 4.2. Tree species diversity relative to plot size in unlogged and logged rain forest, Queensland. Relationships are given as follows: before logging $Hs = 6.2242 \exp(-0.09798 \, \text{AREA}^{-0.48991})$; after logging $Hs = 6.3864 \exp(-0.12923 \, \text{AREA}^{-0.47162})$. (Source: Nicholson *et al.* 1988.)

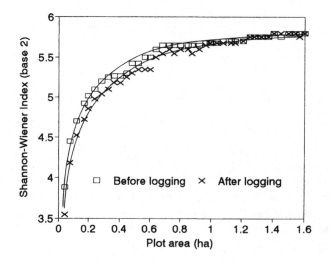

Lianas, for example, are particularly diversified in tropical moist forest regions. In Sabahan dipterocarp forests, densities of liana stems can reach 2000/ha. Most species are phytophilic and respond vigorously to the increased insolation in logged forest. In the Solomon Islands, Whitmore (1984) found more species of large woody climber (50 versus 38/ha) and more individuals (470 versus 194/ha) in forest regenerating on previously cleared land than in primary forest. The abundance of canopy plants such as orchids certainly decreases in logged forest as many host trees are felled and remaining plants are susceptible to decreased humidity. A population of the curious giant-flowered plant parasite *Rafflesia hasseltii* monitored at the Tekam Forest Reserve, peninsular Malaysia, produced 32 flowers in the year prior to logging but no flowering activity was observed in the subsequent year nor when the site was revisited 7 and 13 years after logging[1–13:51]. This was despite the persistence of the host plant *Tetrastigma*.

### Tree regeneration and gap dynamics

*Gap dynamics*

Regeneration of trees in gaps is dependent on three main factors:

1. Gap size. In real terms, the degree of change of microclimatic conditions, which depends on the degree of canopy opening, gap orientation, etc.
2. Gap frequency. Large seedlings have a distinct competitive advantage in a gap over small seedlings. In conditions of frequent gap formation seedlings which grow rapidly after germination are at an advantage over species which germinate but then grow slowly over several years.
3. Timing of gap formation. This is particularly important in dipterocarp forest where the main commercial trees exhibit irregular mass fruiting events and have short-lived seedlings. An axiom of silviculture in dipterocarp forest has been to harvest directly after a fruiting event.

*Gap size*

The diversity of tree species normally found in tropical forests has been correlated with the high frequency of gap formation and the range of gap sizes. Seedlings of different species of tree may achieve optimal growth under different light regimes in different sized gaps. Gaps are formed as part of the normal lifecycle of a tropical forest canopy. Most gaps (around 70% in Sabah; Brown 1990) are formed by the falling of a branch or single tree and do not exceed the diameter of an adult tree crown. In some forest formations, such as those influenced by periodic cyclones, average gap sizes can be larger and gaps occur in distinct cohorts corresponding to the cyclone events.

In broad terms, their are two groups of trees. Climax species are characteristic of primary forest habitats: their seeds will germinate and seedlings become established underneath a closed canopy. Pioneer species are characteristic of open areas and early successions within the forest: germination of their seeds and growth of seedlings generally requires exposure to full radiation. In small gaps, established seedlings of forest trees take advantage of the increased light availability and grow to fill the canopy space available. In large gaps, pioneers which appear only after the gap is formed predominate in the early stages of regeneration. Intuitively, there should be a point at which there is a switch from the growth of established seedlings to the growth of pioneers ('gap switch size') which is expected to be correlated with changes in microclimate associated with large gaps. Identification of switch size has proved elusive and will certainly vary according to forest type. In lower montane forest in Java the switch occurs between 1000 and 2000 m$^2$. In lowland rain forest it is probably less as pioneers can be found in gaps of 80 m$^2$ (Whitmore 1991). However, low levels of pioneer seed germination have been reported to occur in very small gaps and even under a closed canopy (Kennedy & Swaine 1992), such that identifying a gap switch size purely on the basis of presence of pioneers is problematic.

Although fundamental to tropical silviculture it is difficult to assign particular regrowth characteristics to species based on their responses to gaps of particular sizes. Brown (1993) demonstrated that gaps may exhibit considerable temporal and spatial variation in microclimate. There is no simple equation of microclimate with gap size and it is the former to which seedlings are generally responding. Several studies have demonstrated that various climax tree species have the capacity to establish themselves in gaps of a wide range of sizes (Central America: Barton 1984, Brokaw & Scheiner 1989; Queensland: Thompson *et al.* 1988; Sabah: Brown & Whitmore 1992). Within gaps, species may show differences in growth rates at different microsites: at the root plate where soil is disturbed, where the bole fell, or where the crown fell (Brandani *et al.* 1988). For example, fallen crowns may buffer the changes in microclimate in parts of gaps and allow more rapid establishment of climax tree seedlings.

A variety of factors may limit the growth of climax seedlings in large gaps. Brown (1990) found that higher temperatures, causing leaf scorch and wilting, and the incidence of damage by apical shoot borers, limited the growth of seedlings of certain dipterocarp species in large gaps (1500 m$^2$). It nevertheless remains true that pioneer species as a group increase in density with increases in gap size and dominate large gaps (although some species germinate best in smaller and some in larger gaps: Raich & Gong 1990).

*Gap frequency*

In addition to gap size and consequent variation in microclimate, the frequency of gap formation will affect species regeneration. Seedlings germinating under the canopy have differing persistence times. A high frequency of gap formation will favour fast-growing species, but slow-growing species that may persist for a longer time in the understorey will be at an advantage if the frequency of gap formation is very low (Canham 1989). The rotation times of harvesting schedules will clearly influence the frequency of gap formation as well as gap sizes and densities.

*Timing of gap formation*

The seed bank and the densities of seedlings of particular species will fluctuate in size and composition over time. Seeds survive for different lengths of time on the forest floor (a function both of seed longevity and the degree of predation), and seedlings persist for varying lengths of time depending on their growth regime. The trees that eventually become established in gaps in Malaysian dipterocarp forest reflect which seedlings happened to be present when the gap was formed, regardless of its size (Raich & Christensen 1989). The abundance of dipterocarps in such a gap will be largely dependent on the elapsed time since the last mass fruiting of dipterocarps at the point when the gap was formed.

## Tree phenology

Flowering and fruiting peaks of different species may be staggered over the year, although there can be major fruiting seasons broadly related to the seasonality of rainfall. In complex habitats, fruiting peaks may also be staggered between dominant and minor habitat types (Terborgh 1983). The phenology of leafing and fruit production in individual species is keyed into local microclimatic features. Many tree species flower and fruit synchronously in order to maximize pollination success and minimize losses to predators. Fruiting can occur several times per year, annually or biannually, or irregularly. Precisely what triggers the irregular mass fruiting of dipterocarps in South-East Asia remains a mystery (the more so because the New Guinean outlier of the family *Anisoptera thurifera* fruits annually: Whitmore 1990).

Defining causes for phenological phenomena is notoriously difficult and there are few consistent trends that can be attributed to changed environmental conditions subsequent to logging operations. Immediately following logging[1:51] at the Tekam Forest Reserve, peninsular Malaysia, changes in microclimate associated with logging caused a significant increase in leafing activity among residual trees but no significant changes in fruit production (Fig. 4.3). In terms of the total amount of resource per unit area, the volume of leaf material remained constant while fruit availability decreased.

Results from older logged forest are contradictory. At various sites in peninsular Malaysia, leafing activity per unit area has been reported to fall to a level equal to or lower than that of unlogged forest, while fruiting levels are generally higher than in unlogged forest (Chivers 1972; Laidlaw 1994). In logged[11:61] forest at Ponta da Castanha, Brazil, there were no significant changes in leaf production but significantly lower levels of fruit production per unit area (Johns 1991a).

Broad phenological differences may therefore vary according to forest type. Of more importance to frugivores and folivores is the extent to which important food trees are lost during logging and the extent to which pioneer species may be used as alternative food sources. Heydon (1994) demonstrated that although overall differences in fruit availability between unlogged and logged[12:57] forest were not significant, there was a significant reduction in the availability of fruit eaten by ungulates (Fig. 4.4). This reflected a 43% reduction in the density of food plants in logged forest.

Fig. 4.3. Fruit and new leaf production over a felling event, Tekam Forest Reserve, peninsular Malaysia. Figures are percentages of a sample of 1140 trees ≥ 30 cm gbh prior to logging, decreasing to 565 trees thereafter. (Source: Johns 1988.)

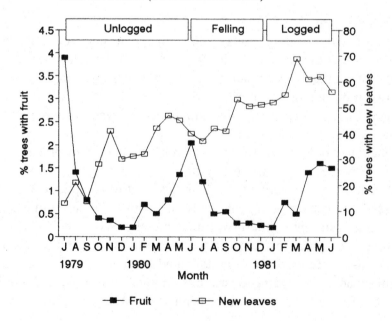

## Management to promote timber tree regeneration

The majority of tropical forest silvicultural systems rely on natural regeneration to produce the next crop of timber, although they differ in reliance on seedlings or larger size classes according to the rotation times envisaged. For economic reasons, replanting of trees is kept to a minimum. The type of regeneration that occurs in forest following logging will be heavily dependent on the disturbance caused by the logging operation.

In most intensive logging operations 50% or more of the seedlings present on the forest floor are destroyed through a combination of tree felling and skidding. Gaps generally exceed switch size and pioneers dominate the regeneration within 6 months (Borhan *et al.* 1987). Climax species may grow through from beneath pioneer trees but the rotation time is lengthened. For this reason, most current logging operations are polycyclic, emphasizing the protection of advanced growth (pole sized trees). This requires active measures to protect marked trees during felling operations. Skilled felling teams should also be able both to preserve advanced growth marked for retention and to control gap size to some extent. However, greatest success is achieved by limiting the number of trees felled and rigid observance of minimum diameters.

It has been suggested that forest management should mimic, or take place within the limits, of a forest's natural disturbance regime (Attiwill 1994a). As a

Fig. 4.4. Mean number of trees ($\geq 30\,\text{cm}\,\text{gbh}$) bearing ripe fruit per 0.4 ha plot ($n = 5$) in unlogged and 12-year-old logged forest, Ulu Segama Forest Reserve, Sabah. Differences are not significant (Friedman test: $s = 0.53$, $df = 1$, $P = 0.47$.) (Source: Heydon 1994.)

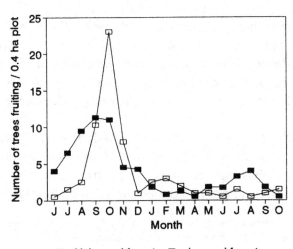

practical example, studies of treefall dynamics and regeneration of logged[11-12:?<35&<50] forest in the Kibale Forest of Uganda have suggested a maximum allowable damage level of 35% which may mimic natural treefall rates of 2.3%/year and thus not significantly influence natural processes (Skorupa & Kasenene 1984).[3]

The management skills exist to control the intensity and frequency of disturbance. It is possible to achieve a defined compromise between maximizing productivity and maximizing biodiversity retention (Attiwill 1994b). Management of forest within natural disturbance regimes would minimize the uncertainty associated with natural resource planning (McCarthy & Burgman 1995), but this generally implies minimal offtake and no intervention. This is not usually economically viable, at least on a large scale. An exception might be logging of forests in the cyclone belt where the high damage levels caused by a typical logging operation mimic the cataclysmic destruction caused by periodic cyclones. (This is perhaps a factor in the apparent success of silvicultural methods formerly practised in the Queensland forests.)

Most commercial hardwoods in tropical forest are climax species (with some exceptions, such as *Swietenia macrophylla* and *Duabanga moluccana*). This implies that disturbance to the canopy during harvesting should be kept to a minimum to reduce average gap size. This minimizes microclimatic change, prevents inhibition of climax species and discourages the widespread growth of pioneers. It is certainly possible to achieve this in practice. For example, a logging operation in Suriname which felled 10 trees/ha achieved an average gap size of $225\,m^2$, resulting in only a slight increase in the proportion of pioneer species in the stand (de Graaf 1986). Of course, the greater the density of timber removed and the greater the damage level, the less likely that gap size can be effectively controlled. In the Kibale Forest of Uganda, natural treefall gaps averaged $256\,m^2$, gaps in forest where basal area was reduced by 15% averaged $467\,m^2$, and gaps in forest where basal area was reduced by 60% averaged $1307\,m^2$ (Kasenene 1987). Regeneration of climax trees was poor in gaps of $>650\,m^2$. Larger gaps were invaded by elephant grass (*Pennisetum*) and tended to grow larger still through windthrow of trees at their edges (Kasenene & Murphy 1991).

Finally, controlling gap size to promote regeneration of desired tree species is, of course, of little value if the seedlings are not present. In logging operations where a high proportion of existing seedlings are destroyed, it may be important to maintain 'seed trees' to provide the necessary stock through the

---

[3] Skorupa & Kasenene's (1984) methodology is an example of the procedure that could be followed, but this precise example cannot be considered accurate since it is based on results drawn from a curve linking three data points.

regeneration process. For example, removal of very high percentages of mature dipterocarps from forests which have not recently mass fruited would result in very low seedling stocks and a probable need to replant (Borhan *et al.* 1987). In the Budongo Forest, Uganda, Plumptre (1995) found that mahogany *Khaya* fruits only at tree sizes > 50 cm diameter and that seedling density correlates strongly with the density of trees over this size. Adult specimens > 50 cm diameter would need to be retained to ensure restocking by natural processes.

### Summary

There is considerable geographical variation in the volumes of timber felled from tropical forests. The level of damage sustained by the forest during logging is broadly correlated with offtake, but is heavily influenced by the extent to which measures for reducing incidental damage are applied. Felling and skidding operations cause most loss of trees, but tree mortality continues into the forest regeneration phase, particularly due to increased windthrow.

Logging in tropical forest is typically non-selective in its overall effects. Large and small trees are destroyed in equal proportions. Although rare species with valuable timbers may be entirely eliminated, logging generally causes a random loss of all tree taxa. Results from small study plots typically show a reduction in species-richness following logging, reflecting overall damage levels, but this is reduced in significance in larger plots due to the random nature of the disturbance. Differences arise in older logged forest due to the growth of a relatively few species of pioneers.

The size and frequency of gaps, and the timing of gap formation, are all important in determining patterns of tree regeneration. While there are few clear relationships between species growth rates and gap size, larger gaps tend to be dominated by pioneers and smaller gaps tend to regenerate climax species more rapidly.

Patterns of leafing and fruiting activity in logged forest are rather variable. In general, animal species dependent on climax rather than pioneer trees for their food will face a reduction in food availability in logged forest.

It has been suggested that tropical forest management should mimic natural regeneration processes. This is usually not economically viable since it implies a very low offtake and no intervention. Natural regeneration dynamics may provide some clues, however, particularly concerning the maximum gap size that will allow efficient establishment of climax species or the required size and density of seed trees for efficient seedling establishment.

# 5

# Responses of individual animal species

### Introduction

Animal communities in tropical forests live within a mosaic of subhabitats caused by different local edaphic conditions and different stages of the lifecycle of treefall gaps. Each patch in the mosaic may be occupied by a different array of plants and animals, some resident and some transient. The degree of transience in animals can be very high (Karr & Freemark 1983). The extent of specialization to particular subhabitats also varies considerably, being most evident among small folivores (such as insect larvae which may live only on certain parts of certain plant species) and least evident among large-bodied frugivore-folivores.

Food resources for frugivores tend to be widely dispersed in the forest, those for folivores less so. Both may be affected by changing patterns of food tree distribution following logging. Many climax food trees are randomly distributed in primary forest, but become patchy following intensive logging operations. In a peninsular Malaysian dipterocarp forest, for example, four important taxa of food trees became localized to the less damaged valleys following a combination of conventional and high-lead logging[1:51] that severely damaged forest along ridgetops (Johns 1988). Such differences may necessitate changes in foraging behaviour among associated animal species. Changes will be most pronounced where the animals are specialists of the trees concerned.

Food resources for insectivores are stable and predictable in unlogged forest but become less so following logging. This reflects changes in the seasonality of primary production. Food resources available to predators and to decomposers may also vary considerably between unlogged and logged forests.

Animals that are specialized to exploit particular food resources that become less common following logging should, logically, also become less common unless they are able to adapt in some way to the changed conditions. Conversely, if trees become too rare to be exploited efficiently by animals, they may lose their pollination and dispersal agents which can affect their long-term viability. The ability of animal species to co-exist alongside tropical forestry operations may be relevant not only to the conservation of that animal species, but to the regeneration and continued functioning of the forest itself.

This chapter reviews existing information on the ability of animal species to co-exist with defined levels of timber logging. Species are first grouped into feeding guilds, but subsequently divided in taxonomic terms. Species assemblages or individual species that provide data of particular interest are considered separately.

### *Limitations of the data*

Currently, very little of the available information results from scientifically rigorous studies. Most analyses lack error terms and samples are frequently too small to be able to apply statistical tests. In most cases (excepting analyses of census data) the relationship between abundance and sighting rates is assumed to be constant between sites: no correcting factors are employed. Perhaps most importantly, most results are derived from comparisons of different sites with different logging histories and thus cannot control for natural spatial heterogeneity in species distributions and abundance (see Chapter 7). These factors limit the degree of certainty associated with conclusions.

### Folivores

Leaves are typically an abundant resource in primary tropical forests although not all leaves are available to folivores. Most animals eat only certain stages (mostly young leaves or shoots, which contain a maximum of usable protein and minimum of lignocellulose and toxic secondary compounds) or specialize on a few or a single species of tree.

### *Invertebrates*

The bulk of leaf material produced in tropical forests is consumed by insects. The amount eaten from each tree varies from a very few shoot tips to complete defoliation (e.g. Whitten & Damanik 1986). The latter is unusual: it is hardly in the herbivore's interest to kill its host tree. A balance is more usual, the abundance of folivorous insects being correlated with seasonality of leaf production. Leaf production typically shows greater annual fluctuation in secondary and logged forests than in primary forests, resulting in equivalent fluctuations in the abundance of foliage insects (Fogden 1972).

In logged[14-17:? < 50] areas of the Kibale Forest of Uganda, the abundance of understorey foliage insects was strongly correlated with the amount of ground vegetation (Nummelin 1989). The amount of available vegetation fluctuated considerably in gaps in logged forest, reflecting the seasonality of rainfall. This correlated (after a time lag) with fluctuations in the density of foliage insects. Similar fluctuations in insect numbers were not recorded in adjacent unlogged forest. In another example, the number of foraging columns of army ants *Eciton* encountered each month in logged[11:61] forest at Ponta da Castanha in Brazilian Amazonia varied from 30–120% of the encounter rate in adjacent unlogged forest (Johns 1986a). This was also considered due to fluctuations in insect abundance in the understorey of logged forest. However, in old logged[23-25:?60] forest at Pasoh, peninsular Malaysia, foliage insects were less abundant overall and periodically a scarce resource in comparison with adjacent unlogged forest (Wong 1985). In this case, the enriched logged forest had regenerated an even-aged stand of commercial timber trees and had few gaps and sparse ground vegetation. The resources for understorey folivorous insects were much reduced.

Within tropical forests, plant species-richness is correlated with the range of microhabitats available. Logged forest may have a reduced number of microhabitats, at least in early regeneration stages, which should also reduce the species-richness of both plants and folivorous insects. Folivorous insects of the upper canopy, such as species associated with epiphytes, orchids, strangling figs or the upper canopy trees themselves, may be particularly adversely affected. Dispersal between host plants may become problematic.

The strong vertical zonation patterns characteristic of primary forest tend to be lost in intensively logged forest. Pioneer vegetation will support new species, such that species-richness regains former levels, but species composition will change. Unlike larger folivores, many insects have co-evolved with specific host plants to the extent that they are unable to revert to other food sources if faced with environmental changes (Holloway & Herbert 1979). Many also have cryptic adaptations which are only effective on specific substrates.

Changes in the diversity of larger nocturnal lepidoptera have been reported as a result of logging, with even more pronounced changes where the forest is cleared temporarily for shifting agriculture or permanently for plantations (Holloway *et al.* 1992). In a short-term trapping programme in Sabah, there was no significant difference in the number of moths caught between unlogged and logged[5:57] forest, but considerable differences in the relative abundance of species and families (Henwood 1986). The greatest degree of difference was among small species. The extent to which adult moths move from their larval habitats is unknown, but probably greater in large bodied species which might thus be expected to show less variability over space. The distribution of larval forms would be most variable between unlogged and logged habitats but no studies have examined this.

In Sarawak, diurnal lepidoptera of the families Satyridae, Riodinidae and Hesperidae were most common in the forest interior, while species of the Papilionidae, Peridae and Danaidae were common in disturbed forest (Holloway 1984a). The latter families occupy the canopy of unlogged forest: the effect of logging is to eliminate stratification of butterfly species and cause the loss of those typical of the forest understorey.

In Neotropical forests, logging causes an initial reduction in the species-richness of the diverse heliconiine butterflies. Species diversity is regained once logged areas are colonized by the rapidly growing *Passiflora* climbers. However, unlogged forest species may be absent in logged forest, being replaced by species characteristic of the forest edge. Similarly, in Uganda total butterfly species-richness did not differ significantly between unlogged (129 species) and various logged[ongoing:? < 50] forests (124–131 species), although there were generally fewer forest interior species and more forest-edge species in the latter habitat (D. Kalibakate, pers. comm.).

Insects with more eclectic diets are generally abundant in old logged forests. These include Neotropical leaf-cutting ants *Atta* and *Azteca*, which exploit a wide variety of plants from canopy trees to planted crops. Another example would be insects which naturally exploit plants of early successional patches, such as Neotropical ithomiine butterflies and other specialists of the Solanaceae. In Papua New Guinea, larvae of birdwing butterflies *Ornithoptera* and *Triodes* are associated with various species of *Aristolochia* climber. These climbers are normally canopy species but are phytophilic and can regenerate rapidly in lower levels of the forest following logging. Birdwings are generally commoner in logged forest and have been harvested subsequent to enrichment planting with specific *Aristolochia* host plants which increase the density of commercially valuable butterfly species (Parsons 1983).

### Smaller vertebrates

Among vertebrates, relatively few species specialize on single or a few plant species. Three-toed sloths *Bradypus* of the Neotropics partition foliage resources through 'clan' systems, groups of animals specializing on different sets of trees and tree species (Montgomery & Sunquist 1978), but this appears to be unique. Potentially at least, sloths are able to eat a wide variety of leaves, including the leaves of pioneers such as *Cecropia*, which are of moderate digestibility but high abundance. In Brazilian Amazonia, particularly high densities of three-toed sloths are recorded in várzea forest which characteristically has a high density of early successional patches. No difference in density was recorded from unlogged and logged[ongoing:6] várzea (Johns 1986a). Both three-toed and two-toed sloths *Choloepus* were found commonly in logged[5–8: < 50] and secondary terra firme forests flooded by the Tucuruí hydroelectric project in eastern Amazonia (Mascarenhas 1985). The latter forest type consisted almost entirely of *Cecropia*. Whether three-toed sloths maintain a clan system of food resource partitioning in secondary forest is unknown.

Obligate folivory becomes more common in large than in small mammals. Only a few small mammals feed entirely on vegetable material (e.g. *Phalanger*, *Capromys*) and these show anatomical or behavioural adaptations, such as coprophagy. Among larger arboreal folivores, colugos *Cynocephalus variegatus* have been recorded from logged[10:51] forests in peninsular Malaysia, but were not observed in logged[6–12:57] forests in Sabah (Johns 1989a). Tree hyraxes *Dendrohyrax arboreus* have been observed in logged[6:? > 50] forests in Sierra Leone (Davies 1987) and occur at high densities in old logged[40–60: > 50] areas of the Budongo Forest, Uganda.

Energetic constraints mean that folivory occurs among very few birds (Morton 1978) and no bats at all. An exception is the hoatzin *Ophiostocomus hoatzin* which possesses an enormous crop for storing and digesting vegetation (Grajal 1995) and remains inactive for long periods. In Brazilian Amazonia these birds are typically observed in mangrove, swamp or riverine vegetation, feeding on young shoots of water-associated plants. Their food plants appear to regenerate rapidly in logged várzea and the population density of hoatzins is not significantly reduced.

While there are no examples of folivorous vertebrates being absent from logged forest areas, some may be susceptible to a reduction in numbers. Studies of folivorous arboreal marsupials in Queensland have shown that one, the lemuroid ringtail possum *Hemibelideus lemuroides*, consistently decreases in density in logged[6–15:?20] rain forest (Laurance & Laurance 1996). This species feeds almost entirely on the leaves of primary forest trees. Four sympatric species feed in part on pioneer tree species or incorporate some fruit into their diet and these do not decrease in density following logging.

The diets of many tropical birds are poorly known and those of large species often contain at least some foliage. The tooth-billed bowerbird *Scenopoeetes dentirostris* of New Guinea not only cuts leaves to construct its mating arena, using its specially-adapted bill for the purpose, but may be totally folivorous for part of the year (Lavery & Grimes 1974). Forest-dwelling tinamous Tinamidae, guineafowl Numididae and partridges Phasianidae may all feed to an appreciable extent on certain vegetation types. Two or three species of tinamou were absent from logged[1-3:6] várzea forest at Mamirauá Lake, Amazonia, although the very low damage levels suggest that this probably reflects spatial heterogeneity in their abundance. In intensive logging operations, damage to the vegetation during logging cannot necessarily be dismissed as a factor influencing their persistence.

While not strictly members of the forest avifauna, aquatic birds such as screamers Anhimidae, some ducks Anatidae and rails Rallidae feed largely on vegetation and may be influenced by changes in water quality occurring after logging.

### Large terrestrial browsers

Since most productivity in primary forests is locked into the canopy, terrestrial folivores are typically browsers rather than grazers and obtain most of their food from treefall gaps or riverine areas where there is a higher density of leaf material close to the ground. Most species travel very long distances within primary forest to find early successional patches. It is common for terrestrial browsers to congregate in logged forests where the density of early successional patches is much greater (Table 5.1). In Sabah, the availability of grass and herbaceous vegetation (in effect the availability of recently logged[1:57] forest) was a significant determinant of the abundance of sambar deer (Heydon 1994).

In the Tai Forest, Côte d'Ivoire, elephant densities increased from 0.5 to 2.6 individuals/km$^2$ following logging[<25:?>50] (Merz 1986). In peninsular Malaysia, elephant densities in logged[1-6:>50] forests were double those of unlogged areas (Olivier 1978). Elephant densities were not significantly different between unlogged and logged[1-15:11] forest in Gabon (White 1992), but in this case logging was of very low intensity and caused relatively little change to the vegetation structure compared with the other examples given.

The high densities of browsers tend to persist in logged forests as long as dense regenerating understorey remains. In peninsular Malaysia, tapirs, which respond rapidly to opening up of the canopy by logging and moved into a studied logging area even before the machines had moved out, were still present at far higher densities in older logged[13-18:51] forests.

## Exudate specialists

Sporadic feeding on exudates flowing from sites of insect or mechanical damage is a fairly common behaviour among small mammals such as squirrels Sciuridae. Exudate specialists are uncommon, consisting of only a few African prosimians, smaller Malagasy lemurs and Neotropical callitrichids. Many, particularly callitrichids, have dental adaptations for gouging their own feeding holes in the bark of suitable trees. One little-known bird, the crested coua *Coua cristata* of Madagascar, feeds largely on *Terminalia* gums.

Exudates very greatly in chemical composition and their exploiters are normally highly selective. In Gabon, prosimians were recorded feeding from only five tree species (all from the family Mimosaceae). An analysis of the gum of the Neotropical tree *Anacardium occidentale*, a common exudate source for marmosets *Callithrix*, found an 84% carbohydrate content and important trace minerals, but no protein (Coimbra-Filho & Mittermeier 1976). It was thus useful primarily as an energy source.

Exudates exploited by the fork-marked dwarf lemur *Phaner furcifer* do

Table 5.1. *Abundances of some terrestrial browsers in unlogged and logged forests*

| Area and species | Signs encountered/100 km surveyed | | | |
|---|---|---|---|---|
| | Unlogged | 1- to 6-year-old logged | 7- to 12-year-old logged | 13- to 18-year-old logged |
| Tekam Forest Reserve (peninsular Malaysia)[a] | | | | |
| Sambar deer *Cervus unicolor* | 0.4 | 3.0 | 6.0 | 1.0 |
| Malayan tapir *Tapirus indica* | 0.7 | 8.0 | 10.0 | 7.0 |
| Seladang *Bos gaurus* | 0 | 1.0 | 2.0 | 0 |
| Asian elephant *Elephas maximus* | 3.0 | 4.0 | 3.0 | 2.0 |
| Lopé (Gabon)[b] | | | | |
| Buffalo *Syncerus caffer* | 12.2 | 2.9 | 6.7 | — |
| African elephant *Loxodonta africana* | 0.5 | 0.2 | 0.2 | — |
| Tucuruí (Brazilian Amazonia)[a] | | | | |
| Brocket deer *Mazama* spp. | 8.0 | 25.0 | — | — |
| South American tapir *Tapirus terrestris* | 4.0 | 8.3 | — | — |
| Ponta da Castanha (Brazilian Amazonia)[a] | | | | |
| Brocket deer *Mazama* spp. | 3.1 | — | 3.1 | — |
| South American tapir *Tapirus terrestris* | 0.2 | — | 2.0 | — |

Figures from Gabon are number of sightings/census (data on the length of census is not available).
All figures are for recent signs of presence, primarily tracks but occasional visual sightings, encountered during diurnal surveys. Survey distances (excluding Gabon) ranged from 40 to 554 km.
*Sources*: [a]A. Grieser Johns (unpublished data); [b]White (1992).

contain protein, reducing the need for this species to supplement its diet with insects (Hladik *et al.* 1980). Coquerel's mouse lemur *Microcebus coquereli*, which inhabits the same forests, obtains its exudates in the form of honey-dew exuded by a flatid bug *Flatida coccinea*. This secondhand exudate has had its protein content reduced and thus the mouse lemur spends a higher proportion of its time foraging for insects.

In Gabon, the needle-clawed bushbaby *Galago elegantulus* occurs commonly in secondary forest, exploiting the largest remaining trees. This is despite its exploitation of fresh exudate droplets rather than congealed masses, and its need to visit between 500 and 1000 feeding sites each night. The fork-marked dwarf lemur is known to occur in degraded forest but may be less common there than in undisturbed forest (Harcourt & Thornback 1990).

Damage to residual trees during logging results in a high availability of flowing exudate in recently logged forest. In the Tekam Forest Reserve, peninsular Malaysia, a variety of species were seen to exploit this temporary resource, including slow lorises *Nycticebus coucang* and many squirrels and flying squirrels. Although the dental morphology of Malaysian squirrels allows species to scrape their own feeding sites, only *Sundasciurus* occasionally does so at times of a fruit and seed shortage. The availability of exudates as an energy source in recently logged forest[1:51] has been hypothesized to be important in maintaining the population of squirrels at a time when there is a severe shortage of fruit and seeds (Johns 1983).

Planting of the exotic *Albizia falcateria* at the Tekam Forest Reserve, primarily to stabilize soils on land heavily damaged by logging, provides a supplementary exudate source in older logged forests. Logged[4–16:51] forest planted with *Albizia* supported 26 individuals/km$^2$ of slow lorises: a similar density to unlogged areas. Populations of slow lorises in logged[8–14:51] forest not planted with *Albizia* declined to around six individuals/km$^2$. Naturally colonizing *Albizia grandibracteata* in disturbed areas of the Kibale Forest, Uganda, were similarly visited by high numbers of galagos *Galago demidovii* and *G. inustus*.

### Pygmy marmosets

The pygmy marmoset *Cebuella pygmaea* of the western Amazon basin spends 67% of its total feeding time (32% of its daily activity) in gouging exudate holes and feeding from them (Ramirez *et al.* 1977). Groups normally restrict their activity to the vicinity of a large emergent exudate tree (*Vochysia*, *Spondias*, *Parkia* or *Qualea*) until the flow of exudates diminishes, whereupon the group decamps to a known source tree nearby or emigrates in search of another. The species occupies a succession of small but widely spaced home ranges, which appears to be an appropriate strategy for exploitation of the widely dispersed but concentrated exudate resource.

The species's ability to migrate between food sources makes it a successful colonist of even highly disturbed forest (Soini 1982). Although some food trees, notably *Parkia*, are commercial timbers the species appears adept at locating residual trees in both logged forest and areas of shifting cultivation. Their population density can be as high as 50 individuals/km$^2$ in both undisturbed and disturbed forests.

### Frugivores and seed predators

Patterns of fruit production vary a great deal between and within tropical forests. Dipterocarp forests, for example, possess many fewer tree species producing bird-edible fruit than African or Neotropical forests. Species of understorey tree and those of early successional habitats generally produce smaller sizes of fruit than do upper canopy trees (Foster & Janson 1985).

Many fruit trees occur at low densities in tropical forest, often several kilometres apart in the case of individuals of some species of fig tree *Ficus*. Most frugivorous species travel long distances in the course of their daily activities and form cohesive groups or flocks. Most frugivores, especially those feeding on ripe fruit pulp, are able to utilize a wide range of fruit sizes. Since edible fruit are rarely cryptic, necessitating particular foraging strategies, segregation among frugivores has been mostly by size. A few species, notably seed predators, have developed morphological specializations for feeding on particular under-utilized fruits.

Specialization of animals to exploit a resource that is both widely distributed and erratic in its seasonality is to some extent pre-adaptive to their persistence in disturbed forest. The density of large fruit sources is reduced in logged forest but most frugivorous species are physiologically and anatomically adapted for extensive ranging. This is also true of most facultative seed predators, such as Neotropical macaws *Ara*. The black colobus *Colobus satanas*, a seed predator of central African forests, may operate close to ecological limits in parts of its range (McKey & Waterman 1982). In marginal habitats they are unable to increase range sizes, or distances travelled daily, to compensate for losses of food trees with logging. In less marginal parts of their range they are able to adapt accordingly (White 1992).

Less wide-ranging species which feed on sugar-rich fruits are often able to exploit colonizing shrubs and climbers (Fogden 1972). Lipid-rich fruits are rarer among colonizing plants (*Inga* of the Neotropics being an exception) and small-bodied specialists of these fruit, such as the green broadbill *Calyptomena viridis* of Malaysia (Lambert 1990) may be adversely affected by logging.

### Uakaris and bearded sakis

Two groups of Neotropical pithecine monkeys, bearded sakis *Chiropotes* and uakaris *Cacajao*, are specialist predators of seeds from large unripe fruits (Ayres 1981, 1986b). Both have specialized jaw musculature to enable exploitation of hard-shelled fruit inaccessible to other animals, such as those of the brazil-nut family Lecythidaceae.

Both bearded sakis and uakaris associate in large groups. Bearded sakis move rapidly between fruit trees, foraging as a single unit. Uakari groups tend to split into foraging subgroups which disperse along the forested restingas of their várzea habitat. The linear nature of the restingas makes relocation of other subgroups relatively easy. The uakaris foraging behaviour adjusts the size of subgroups to suit the size of food resources, and probably makes uakaris less sensitive to habitat disturbance than bearded sakis.

Some of the main large fruit trees exploited by southern bearded sakis *C. s. satanas* are also important timber trees (*Manilkara huberi, Eperua bijuga, Pouteria* sp.). With no controls on logging activity, which is generally the case in the eastern Amazon, these species can be eradicated. However, individuals occur only at very low densities, such that even cutting of every mature specimen may result in quite low damage levels. Logging[1–2:10] of only 1–2 timber trees/ha in the Gorupí Forest Reserve reduced bearded saki density by 50% within 2 years (Table 5.2). Bearded sakis are apparently unable to survive

Table 5.2. *Densities of some pithecine seed predators in unlogged and logged forest, Brazilian Amazonia*

| Species and area | Individuals/km² | | |
| --- | --- | --- | --- |
| | Unlogged | Light logging level (≤ 15% of trees destroyed) | Moderate logging level (≥ 50% of trees destroyed) |
| Southern bearded saki *Chiropotes satanas* | | | |
| Tucuruí[a] | 9 | — | 0 |
| Gorupí Forest Reserve[a] | 30 | 15 | 0.2 |
| White-nosed saki *Chiropotes albinasus* | | | |
| Tapajós National Park[b] | 8 | 13 | — |
| White uakari *Cacajao calvus calvus* | | | |
| Mamirauá Lake[a] | 16 | 24 | <10 |

*Sources*: [a]Johns (1986a); [b]Reanalysed data from Branch (1983).

in more intensively logged forest both at Gorupí[(4:50)] and Tucuruí[(6:>50)], perhaps because even low-quality substitute food sources are lost. An apparently anomalous result from the Tapajós National Park, where white-nosed sakis *C. albinasus* were actually commoner in logged forest[(?:<15)], may be explained by the facts that Branch (1983) surveyed floristically dissimilar forests and that logging targeted only *Aniba duckei* which does not provide a food resource for sakis.

Uakaris appear less affected by light[(ongoing:6)] and moderate[(ongoing:50)] logging than do bearded sakis, despite similar adaptations to feeding on rare and widely dispersed fruit sources and the loss of some of these trees as commercial timbers (Ayres & Johns 1987). Flexibility in foraging may help persistence where the number of fruit trees is reduced, as may certain adaptations to seasonal food shortages such as terrestrial foraging for old fallen seeds (Ayres 1986b). Most current logging operations in várzea are not very damaging, but where heavy logging levels are reported (in this case, in an area close to a seasonally inhabited village) declines in uakari density are apparent.

### Other frugivorous primates

Tropical forest primate communities can contain up to about 14 sympatric species, with western Amazonia and eastern Zaire the most species-rich. Within large arrays, several species are predominantly frugivorous. In the Neotropics these are the owl monkeys *Aotus*, sakis *Pithecia* and spider monkeys *Ateles*. The former two may feed on insects or foliage on occasion, but *Ateles* is rarely observed feeding on anything except fruit and flowers. *Ateles* tends to be scarce in logged forests, an attribute it shares with the equally large-bodied but largely folivorous woolly monkey *Lagothrix*. These species may experience difficulties in locomotion through heavily damaged forest, where there are fewer large substrata able to bear their weight.

Perhaps equally significant for *Ateles* is its inability to exploit the small dispersed fruit sources of the types used by *Aotus* and *Pithecia*. At Ponta da Castanha in the western Brazilian Amazon, both *Aotus nigriceps* and *Pithecia albicans* were commoner in logged[(11:61)] forest than in adjacent unlogged forest (18 versus 9, 22 versus 8.5 individuals/km$^2$, respectively: Johns 1991a). Mean group size of *Pithecia albicans* decreased from 5.2 in unlogged to 3.9 in logged forest. Groups split more often in logged forest in order to exploit climbers and pioneer trees, notably *Inga*, which produce an abundance of small fruit through most of the year. At Tucuruí in the eastern Amazon, *Aotus* were also commonly observed feeding on soft fruits from climbers in forests disturbed by shifting cultivation.

In Ghana, the largely frugivorous red colobus and diana monkeys *Colobus badius* and *Cercopithecus diana* are significantly reduced in numbers in logged[(ongoing:50)] forest compared with sympatric generalist species. A large part of the diet of *C. badius* in primary forest consists of the fruits of commercial timber species (Table 5.3), and they are unable to use colonizing trees as food sources. In the Kibale Forest, Uganda, the population density of the red-tailed monkey *Cercopithecus ascanius* has shown a measure of co-variance with fig tree density (Skorupa 1986). In an area of logged[(12:?<50)] forest where fig abundance declined from 4.1 to 0.6 trees/ha, the density of *C. ascanius* was reduced by 50%. Overall tree diversity has been shown to decrease following logging[(8-9:<50)] in dry deciduous forest in Madagascar (Ganzhorn *et al.* 1990), and since the abundance of frugivorous lemurs correlates with tree diversity (Hawkins *et al.* 1990) this is likely to have long-term deleterious effects.

In Sabah, orang-utans *Pongo pygmaeus* have been reported to avoid active logging areas by some observers (Davies 1986) but not by others (Payne 1988). Intensive surveys have determined that orang-utan density in logged[(1-12:>50)] and unlogged forests is not significantly different (Payne 1988; Johns 1989a), suggesting that factors other than forest structure and fruit availability limit their density. By contrast, the chimpanzee *Pan troglodytes* occurs at significantly lower densities in comparatively recently logged[(3-15:>50)] forests (Skorupa 1986; White 1992), but not in old logged[(40-60:>50)] forests where the availability of preferred foods has increased (Plumptre & Reynolds 1994). It has been hypothesized that a drop in chimpanzee density in recently logged[(1:11)] forest in Gabon is due not to resource limitations directly but to inter-group conflicts and associated mortality (White & Tutin in press). Chimpanzees are forced to range more widely in logged forests and in the Gabonese study violent encounters were thought to result. Other studies, such as that in Budongo

Table 5.3. *The use of commercially important trees as food sources by Ghanaian red colobus monkeys*

| Economic class | No. tree species in study area | No. species exploited by red colobus | Recorded food items in diet (%) |
| --- | --- | --- | --- |
| Main timber trees | 11 | 7 | 17 |
| Minor timber trees | 4 | 3 | 5 |
| Trees cut only on rare occasions | 18 | 18 | 43 |
| Non-commercial trees | 127 | 23 | 35 |

*Source*: Adapted from Asibey (1978).

Forest, Uganda, provide no evidence of increased violent encounters although there is overlap of male territories in logged forest (V. Reynolds, pers. comm.).

### Squirrels

A geographical difference may exist in the organization of squirrel (Sciuridae) species arrays. The diversity of African forest squirrels has been correlated with plant species diversity and the predictability of resources on an annual basis (Emmons 1980). Throughout the year species are segregated by foraging height and body size. In peninsular Malaysian dipterocarp forest segregation is seen when fruit is abundant, but this peak fruit crop is not predictable and can only be exploited by species able to subsist on alternative foods for most of the year. At times of low fruit availability a high dietary overlap occurs, with all species feeding on a few common fruiting trees and alternative foods, such as exudates (Johns 1983). These differences are probably due to the low density of trees producing edible fruit in dipterocarp forest.

Since the African species are more specialized, they may be more seriously affected by logging operations which affect the diversity and predictability of food resources. A few African species appear to persist poorly in secondary and logged[6:750] forest (the large-bodied terrestrial species *Epixerus ebii* and semi-terrestrial *Funisciurus pyrrhopus* and *F. lemniseatus*: Emmons 1980; Davies 1987). This has been attributed to changed food abundance and competition with other taxa. Changes in abundance do not follow clear patterns, however, perhaps because squirrels have a very conservative morphology and may have a capacity to adapt to changes in the availability of different food type.

Responses of peninsular Malaysian squirrels to logging[1-18:51] also do not follow clear patterns (Table 5.4). There appears to be a reduction in the population density of some species, but data are of insufficient quality to determine whether inter-specific competition actually results in the elimination of species. There is no equivalent to the large terrestrial *Epixerus ebii* in peninsular Malaysia, but populations of the small terrestrial species *Lariscus insignis* may be declining. In Sabah, the giant tufted ground squirrel *Reithroscuirus macrotis* is rarely observed outside of unlogged forest. Terrestrial squirrel species seem least able to adapt to conditions in logged forest, although most can climb into the understorey to reach available food resources if necessary.

### Frugivorous bats

The species-richness of frugivorous bats tends to decline in logged forests or other disturbed areas. In peninsular Malaysia, logged[1:51] forest is utilized mostly by a few species of pteropodid fruit bat (principally *Cynopterus brachyotis*). In Sabah, four species of pteropodid fruit bat (*C. brachyotis, Balionycteris maculata, Megaerops ecaudatus* and *Aethalops alecto*) were trapped in unlogged forest but only the first two species in logged[6:57] forest. The capture rate of frugivorous species dropped from 5.5/100 net hours in unlogged to 3.0/100 net hours in logged[6:57] forest. Bats of the pteropodid genera *Epomorphus* and *Rousettus* tend to persist in disturbed forests in West Africa and the Indian subcontinent, but abundance also declines.

Neotropical bat faunas have larger numbers of frugivorous and mainly frugivorous species. A secondary forest in Panama contained a slightly lesser species richness of frugivorous species and a tendency for certain species to dominate (Table 5.5).

Table 5.4. *Abundances of diurnal squirrels recorded from line transect surveys in unlogged and logged dipterocarp forest at Tekam Forest Reserve, peninsular Malaysia*

| | | No. individuals/100 km | | |
| | | 1- to 6-year-old logged[a] | 7- to 12-year-old logged[b] | 13- to 18-year-old logged[b] |
| Species | Unlogged | | | |
|---|---|---|---|---|
| *Ratufa bicolor* | 30 | 10 | 9 | 48 |
| *R. affinis*[c] | — | 3 | 5 | 8 |
| *Callosciurus prevostii*[c] | — | 7 | 5 | 0 |
| *C. notatus* | 17 | 20 | 47 | 83 |
| *C. caniceps* | 4 | 0 | 9 | 13 |
| *C. nigrovittatus* | 8 | 4 | 6 | 10 |
| *Sundasciurus hippurus* | 4 | 16 | 8 | 5 |
| *S. tenuis* | 36 | 12 | 15 | 95 |
| *S. lowii* | 3 | 3 | 3 | 3 |
| *Lariscus insignis* | 12 | 18 | 6 | 0 |

Squirrel abundances in all three categories of logged forest were significantly different to abundances in unlogged forest ($\chi^2$ tests: $P > 0.05$; the two species with disjunct distributions were excluded).
[a]Data from three sites combined.
[b]Data from four sites combined.
[c]These species had a patchy distribution at Tekam and did not occur in the unlogged forest site.
*Source*: A. Grieser Johns (unpublished data).

*Hornbills*

The majority of hornbills are tropical forest birds although a few African genera have adapted to open woodland habitats (*Tockus, Bucorvus*). Up to nine species occur sympatrically in Malaysian dipterocarp forest and eight species in eastern Zaire rain forest. This is possible due to co-occurrence of territorial and nomadic species and some divergence in size and dietary specialization. The unusual breeding habits of hornbills, with specific requirements for large tree holes, led to their being among the first species considered likely to be threatened by timber logging (McClure 1968).

Among Asian forest hornbills, which are better studied than African species, three types of fruits are exploited: lipid-rich capsular fruits such as *Aglaia* and *Myristica*, lipid-rich drupaceous fruits of the Lauraceae and Annonaceae, and sugar-rich figs *Ficus*. Figs have been considered particularly important for territorial hornbills (all Asian genera apart from *Rhyticeros*) (Leighton & Leighton 1983). Although not themselves timber trees, many fig trees are destroyed during logging. The host trees of strangling varieties are frequently canopy timber trees. A 74% loss rate of fig trees was recorded following logging[1:51] at the Tekam Forest Reserve in peninsular Malaysia, compared with an average loss of hornbill food trees of 56% (Table 5.6). Pioneer trees do not generally provide fruit eaten by hornbills, although the sugary fruits of some rapidly regrowing climbers are eaten by *Anthracoceros*. The loss of high

Table 5.5. *Catches of frugivorous and largely frugivorous bats in matched mist-netting samples in undisturbed and secondary forest at Gatún Lake, Panama*

| Species | No. individuals trapped | |
|---|---|---|
| | Undisturbed forest (Bohio Peninsula) | Secondary forest (Buena Vista) |
| *Phyllostomus discolor* | 0 | 2 |
| *Carollia castanea* | 33 | 23 |
| *C. perspicillata* | 9 | 73 |
| *Uroderma bilobatum* | 2 | 12 |
| *Vampyrops helleri* | 1 | 1 |
| *Vampyressa pusilla* | 9 | 0 |
| *V. major* | 4 | 0 |
| *Chiroderma villosum* | 1 | 0 |
| *Artibeus jamaicensis* }<br>*A. lituratus* } | 37 | 11 |

*Source*: Johns *et al.* (1985).

canopy trees also reduces foraging substrata, such as loose bark and epiphytes, which are probed by large hornbill species looking for animal prey which provide important protein and minerals in their diet.

Despite the reduction in food tree density, hornbill populations are not noticeably reduced in logged forests in Malaysia (Table 5.7). Persistence of all species of hornbills has also been reported in logged[4: > 50] forest in Sarawak (Kemp & Kemp 1975). Hornbills need to travel further between food sources in logged forest, but they are very mobile animals. Since there is a lesser density of fruiting trees, the number of individual hornbills visiting each tree increases. As many of these large trees produce a superabundance of fruit, this does not appear to result in feeding competition. For species which feed mainly from

Table 5.6. *Loss rates of some hornbill food trees during logging at Tekam Forest Reserve, peninsular Malaysia*

| Family | Genus and species | Trees/ha[a] | | Loss rate (%) |
|---|---|---|---|---|
| | | Unlogged forest | After logging (51% loss of trees) | |
| Annonaceae | *Xylopia* spp. | 6.7 | 4.9 | 29 |
| Burseraceae | *Canarium* spp. | 6.3 | 2.9 | 54 |
| | *Dacryodes* spp. | 2.4 | 0 | 100 |
| Ebenaceae | *Diospyros* spp. | 11.5 | 4.8 | 58 |
| Euphorbiaceae | *Baccaurea* spp. | 10.0 | 6.3 | 37 |
| Fagaceae | *Castanopsis curtisii* | 4.8 | 3.8 | 21 |
| | *Lithocarpus* spp. | 3.4 | 1.4 | 57 |
| Lauraceae | *Alseodaphne* spp. | 2.9 | 1.0 | 66 |
| | *Litsea* spp. | 12.0 | 6.7 | 44 |
| Leguminosae | *Milletia atropurpurea* | 4.3 | 1.0 | 66 |
| Meliaceae | *Aglaia* spp. | 9.6 | 4.3 | 55 |
| | *Amoora* spp. | 1.0 | 0 | 100 |
| | *Chisocheton* spp. | 8.7 | 4.8 | 45 |
| | *Dysoxylum* spp. | 7.7 | 2.9 | 62 |
| Moraceae | *Ficus* spp. | 1.9 | 0.5 | 74 |
| Myristicaceae | *Knema* spp. | 6.7 | 2.4 | 64 |
| | *Myristica* spp. | 5.3 | 3.4 | 36 |
| Myrtaceae | *Eugenia* spp. | 12.5 | 8.7 | 30 |
| Mean loss rate | | | | 56 |

[a]Includes all trees $\geq 30$ cm girth: not all would be mature.
*Source*: Johns (1987).

small fruiting trees the foraging group size declines (e.g. mean group size of *Annorhinus galeritus* in Ulu Segama, Sabah, declined from 7.0 in unlogged to 3.7 in logged[6–12:57] forests).

However, there is a point at which food resources are reduced sufficiently to preclude the continued use of the forest by hornbills. Burning of extensive areas of logged forests in East Kalimantan as a result of the 1982–83 El Niño/Southern Oscillation event caused all hornbill species to migrate from the area, at least on a temporary basis (M. Leighton, in Kemp 1985).

While the availability of food may not be a limiting factor in logged forests, a reduction of breeding sites may be. Hornbills are long-lived birds (Kemp 1995) and a reduction in breeding success might take decades to become expressed in terms of population densities. The potential importance of maintaining breeding sites will be discussed in chapter 8.

### Other frugivorous birds

Few frugivorous birds are intolerant of microclimatic fluctuations in logged forest, unlike insectivores (see below). Most species range throughout

Table 5.7. *Changes in Malaysian hornbill density in logged dipterocarp forest*

| | | Individuals/km² | | | | | |
|---|---|---|---|---|---|---|---|
| | | Tekam Forest Reserve, peninsular Malaysia | | | | Ulu Segama Forest Reserve, Sabah | |
| Species | Unlogged | 1- to 6-year-old logged[a] | 7- to 12-year-old logged[b] | 13- to 18-year-old logged[b] | Unlogged[c] | 6-year-old logged[c] | 12-year-old logged[c] |
|---|---|---|---|---|---|---|---|
| *Berenicornis comatus* | 0.3 | p | 0.1 | 0.1 | 0.1 | 0 | p |
| *Rhyticeros undulatus* | 1.4 | 1.7 | 1.1 | 3.2 | 5.4 | 5.4 | 3.5 |
| *R. corrugatus* | 0 | 0.4 | 1.3 | 0.9 | 0.1 | 0.4 | 0.5 |
| *Annorhinus galeritus* | 2.0 | 2.0 | 1.2 | 2.7 | 3.0 | 2.9 | 0.8 |
| *Anthracoceros malayanus* | 0 | 0.2 | 0.5 | 1.0 | 0.3 | 1.6 | 2.5 |
| *A. albirostris* | 0 | 0 | p | 0 | 0 | 0 | 0 |
| *Rhinoplax vigil* | 2.6 | 2.1 | 1.1 | 1.7 | 1.4 | 2.5 | 0.3 |
| *Buceros rhinoceros* | 3.2 | 2.4 | 3.2 | 5.2 | 2.3 | 3.8 | 2.7 |
| *B. bicornis* | p | 0 | p | 0.1 | — | — | — |
| Indet. | | | 0.2 | 0.6 | 0.6 | | |
| Total hornbill density | 9.5 | 8.8 | 8.7 | 15.5 | 13.2 | 16.6 | 10.3 |

Data are derived from line transect surveys of 30–180 km. Not all hornbills encountered during surveys were identified to species: these are recorded above as Indet. Tree losses during logging were 51% at Tekam and 57% at Ulu Segama.

p, Present.

[a]Data from three sites combined.

[b]Data from four sites combined.

[c]Data from two sites combined.

*Source*: A. Grieser Johns (unpublished data).

the vertical layers of the forest and between mature forest and successional patches, often in large flocks (e.g. parakeets *Pyrrhura* and toucans *Ramphastos* in Amazonia: Johns 1991b). Flocks in logged forest are often larger than in unlogged areas. Flocks of up to 30 scarlet macaws *Ara macao* visited logged[11:61] forest at Ponta da Castanha, Brazil, and flocks of several hundred green pigeons *Treron capelli* and *T. curvirostris* have been seen in fruiting fig trees in logged[6:57] forest in south-east Sabah.

In areas where deforestation has taken place, flocks of some frugivores may persist by moving between such fruit trees as remain in the residual forests (e.g. golden parakeets *Aratinga guarouba* in the eastern Amazon basin, Johns 1986a; subtropical rain forest pigeons in southern Australia, Date *et al.* 1991). There is a limit of food reduction at which point frugivores begin to be eliminated, however, as has been reported in the extensively deforested Brazilian Atlantic region (Willis 1979).

A variety of small-bodied understorey frugivores, such as flowerpeckers *Prionochilus* and *Dicaeum* in Malaysia and manakins *Pipra* in the Neotropics, occur at lesser abundances in recently logged forests. These birds feed from small fruit sources, including trees of the Melastomataceae, Rubiaceae and Magnoliaceae, small-fruited figs *Ficus* and mistletoes Loranthaceae. Many of these understorey trees are destroyed during logging and some, such as figs and mistletoes, which require host trees, may regenerate very slowly. The abundance of flowerpeckers was greatly reduced following logging[1-6:51] in peninsular Malaysia, but recovered following establishment of the pioneer *Melastoma* and other berry-producing shrubs which colonize the verges of logging roads (Table 5.8). Similarly, manakins appeared to be out-competed by tanagers Thraupidae in recently logged[1-4:≤50] forest but regain their numbers in older logged[11:61] forest once an understory has been re-established (Johns 1986a, 1991b).

There are few seedeaters among tropical forest birds although some do occur in secondary forest habitats. In Malaysia, munias *Lonchura* frequent large treefall gaps or landslide areas within primary forest and can become common in heavily damaged logged forest areas colonized by grasses. In the Tabin Forest Reserve of Sabah, munias made up 4.7% of birds encountered during surveys in logged[14&9:32] forest. However, at Tekam Forest Reserve, peninsular Malaysia, they never exceeded 0.6% of samples in logged[1-18:51] forest.

As indicated above, there is considerable local variation in the extent to which grasses colonize logged forest. A high abundance of seedeaters can in some circumstances be correlated with changes in forest succession due to local climatic change. In Pará and Maranhão states, Brazil, northward encroachment of dry cerrado vegetation into formerly forested areas has led to invasion by seedeating finches such as *Sporophila* spp. and *Oryzoborus maximilliani* (Johns 1986a).

### Nectarivores

Nectarivores demonstrate many of the same feeding strategies as frugivores, since their food resources are equally widely distributed and irregular in their seasonality. Few animals feed entirely upon nectar. Many nectarivorous insects also obtain nutrients from pollen, bruised or rotten fruit, soil or animal faeces, and most nectarivorous birds are least partly insectivorous. Co-evolution between plants and nectarivores is nonetheless common, with the animals acting as pollination agents. In some cases, as in the Orchidaceae, specialization to particular insect pollinators can be extreme. Some birds are adapted to feed from locally distributed plants with unique corolla shapes, such as the sword-billed hummingbirds *Ensifera* and *Eutoxeres* of the Andes and sugarbirds *Promerops* of southern Africa.

Most nectarivorous species are not specialized to particular food plants, however. Neotropical euglossine bees travel among, and pollinate, a large number of flowering plants of early successional patches (Apocynaceae,

Table 5.8. *Changes in the abundances of flowerpeckers Dicaeidae following logging at Tekam Forest Reserve, peninsular Malaysia*

| Species | Unlogged | No. individuals/100 km surveyed | | | |
|---|---|---|---|---|---|
| | | 0- to 6-month-old logged | 1- to 6-year-old logged[a] | 7- to 12-year-old logged[b] | 13- to 18-year-old logged[b] |
| *Dicaeum chrysorrheum* | 0 | 0 | 0 | 0 | 3 |
| *D. concolor* | 20 | 0 | 0 | 13 | 7 |
| *D. trigonostigma* | 3 | 0 | 0 | 0 | 0 |
| *D. agile* | 0 | 0 | 0 | 3 | 0 |
| *Prionochilus thoracicus* | 3 | 0 | 0 | 0 | 0 |
| *P. maculatus* | 3 | 0 | 0 | 45 | 18 |
| *P. percussus* | 15 | 6 | 0 | 10 | 13 |
| Total bird sample represented by Dicaeidae (%) | 0.48 | 0.07 | 0 | 0.78 | 1.20 |

[a]Data from three sites combined.
[b]Data from four sites combined.
*Source*: A. Grieser Johns (unpublished data).

Marantaceae, Rubiaceae), the understorey (Marantaceae, Rubiaceae) and the canopy (Orchidaceae, Bignoniaceae, Bromeliaceae). The maintenance of a diverse euglossine bee fauna relies on a diversity of plants from a variety of successional stages, some being larval foodplants, some nectar sources, and some sources of volatile compounds used in courtship behaviour. Although extensive deforestation results in the loss of some euglossine bee species (Lovejoy *et al.* 1986) this might not be expected following logging. Many larval and adult foodplants are early successional plants abundant in logged forest (e.g. Convulvulaceae, Passifloraceae). The only potential problem for the bees would be an inability to locate sources of chemical compounds used in courtship, many of which come from canopy epiphytes.

Neotropical callitrichids, notably *Saguinus* may be heavily reliant upon nectar during certain seasons (Terborgh 1983). A number of bats may be largely nectarivorous (*Glossophaga* of the Neotropics, *Eonycteris* of Asia, *Micropteropus* of Africa). Nectarivory is most common in birds, however. Several groups of birds are specialist nectarivores, although most obtain additional proteins from insects. These include the hummingbirds Trochilidae and flower-piercers *Diglossa* of the Neotropics, sunbirds Nectariniidae of Africa and Asia, and honeyeaters Meliphigidae of New Guinea.

In Amazonia, emerald-type hummingbirds of genera such as *Amazilia*, *Chlorestes* and *Chlorostilbon* are usually common in logged forests. Available data (Table 5.9) suggest a drop in abundance and perhaps in species richness, but this is anomalous and perhaps an expression of changes in social relationships (see below). The predominantly insectivorous hermits *Phaethornis* and *Threnetes* are certainly less abundant in logged forests. These species specialize in picking insects from spiderwebs, which are a more predictable resource in primary forest. In Malaysia, sunbirds of the genera *Anthreptes* and *Hypogramma* occur commonly in logged forest (Table 5.9) visiting flowering early successional plants. The flower-piercers *Diglossa* have bills adapted to pierce the corolla tubes of flowers. An ability to break into a resource regardless of corolla shape is advantageous in logged forest where the abundance of different flower types changes.

### Social organization in hummingbirds

In mature Neotropical forests, flowers pollinated by hummingbirds tend to be widely divergent in shape. Hummingbirds have evolved divergent bill sizes to exploit particular flowers. Flowering plants typical of early successions produce corollas with a generalized shape. Potentially these could be exploited by all or most hummingbird species, even the shortest-billed.

In primary forest, hummingbird species feed separately, employing their specializations for flowers of different morphologies. However, most persist in disturbed habitats even if some groups of nectar-producing plants are absent. There are two possible reasons for this. First, few species are obligate associates of particular flowers although they may have particular specializations to allow their exploitation (specializations are an expression of competitive relationships). Second, nectar-producing shrubs and climbers are very common in early successional habitats, which buffers the loss of canopy orchids and other flowering epiphytes.

The abundance of hummingbirds in secondary forests varies seasonally according to the spatial dispersion of flowering among nectar source plants. Local abundance of particular species is also partly determined by aggressive defence of resources by a few species at times of low nectar availability (Feinsinger 1976). Certain species may be excluded from available nectar sources, causing considerable local movements of these birds. Since hummingbirds appear able to move quite long distances in search of available nectar sources, avoiding defended sources if necessary, competition need not eliminate species from logged areas. This may, however, be a factor in the drop in

Table 5.9. *Species-richness and abundance of nectarivorous hummingbirds and sunbirds in unlogged and logged forests of Brazil and Malaysia*

| Species and study site | Unlogged | | 1- to 6-year-old logged | | 7- to 12-year-old logged | | 13- to 18-year-old logged | |
|---|---|---|---|---|---|---|---|---|
| | No. species | No. indiv. per 100 km | No. species | No. indiv. per 100 km | No. species | No. indiv. per 100 km | No. species | No. indiv. per 100 km |
| Trochilidae[a] | | | | | | | | |
| Gorupí Forest Reserve | 4 | 45 | | | | | | |
| 10% tree loss | | | 2 | 22 | — | — | — | — |
| 50% tree loss | | | 2 | 20 | — | — | — | — |
| Ponta da Castanha | 4 | 36 | | | | | | |
| 60% tree loss | | | — | — | 4 | 29 | — | — |
| Nectariniidae[b] | | | | | | | | |
| Tekam Forest Reserve | 4 | 85 | | | | | | |
| 51% tree loss | | | 4[c] | 57 | 3[d] | 74 | 4[d] | 73 |
| Ulu Segama Forest Reserve | 4[e] | 22 | | | | | | |
| 57% tree loss | | | 6[e] | 67 | 6[e] | 115 | — | — |

[a]Excludes the predominantly insectivorous *Threnetes* and *Phaethornis*.
[b]Excludes the predominantly insectivorous *Arachnothera* spp.
[c]Data from three sites combined.
[d]Data from four sites combined.
[e]Data from two sites combined.
*Source*: A. Grieser Johns (unpublished data).

hummingbird species richness recorded where the forest is reduced to small 'islands' between which individual birds may not be able to move (Willis 1979).

### Frugivore-folivores

An ability to adapt feeding strategies to the relative abundance of different resources available is an important feature promoting the survival of animal species in logged areas, particularly during the first few years after logging. The ability to shift between frugivory and folivory is characteristic of many large-bodied mammals. Small-bodied mammals, and also birds, typically shift between frugivory and insectivory (Chivers & Hladik 1980).

#### Primates

A primate species's degree of frugivory shows a significant negative correlation with its ability to persist in recently logged forests (expressed by comparative population densities in unlogged and adjacent logged forests), even at a univariate level of analysis (Johns & Skorupa 1987). Where the effects of confounding variables are controlled, the correlation becomes remarkably strong. Dietary diversity, on the other hand, does not appear to affect a species's ability to persist following logging. In most tropical forests, typical logging operations do not appear to change the diversity of potential food types, although this might be expected where tree loss rates become extremely high. The exceptions are climax forests dominated by one tree species, within which logging or other disturbance normally increases tree diversity and can result in higher primate populations (as in the *Gilbertiodendron* forests of eastern Zaire and the *Cynometra* forests of western Uganda: Thomas 1991; Plumptre *et al.* 1994).

The most important factor affecting a species's persistence is an ability to change the relative proportions of different food types in the diet, specifically to exploit available new leaves in the absence of fruit. Highly specialized frugivores are less able to do this. The most successful species are those which can survive on a largely folivorous diet, even if they are behaviourally frugivorous in primary forest. For example, in unlogged forest at Ponta da Castanha, Brazil, red howler monkeys *Alouatta seniculus* specialized on soft fruits such as figs, while in adjacent logged[11:61] and regenerating cleared forest other groups fed primarily on young leaves of pioneer tree species.

In the Kibale Forest, Uganda, the biomass of largely frugivorous primates was reduced by 59% in a compartment where tree basal area declined by 60% subsequent to logging[12:? < 50]. The biomass of largely folivorous primates was 39% less than in nearby unlogged forest. While it proved difficult to identify particular botanical variables closely correlated to densities of primate species,

Skorupa (1986) hypothesized that the densities of primate species most closely associated with mature forest should show strong positive co-variance between primary and logged forest plots. Analysis by an inter-specific correlation matrix identified four species as likely to be adversely affected by disturbance to mature forest (Fig. 5.1). All four are mostly frugivorous and show less flexibility in diet than other species present (with the exception of *Colobus guereza* which feeds mainly on young leaves from a few pioneer tree species). At the Budongo Forest, Uganda, three of the four species not adversely affected by logging in Kibale occur at higher densities in logged[(1-60:> 50)] than in unlogged forest (Plumptre *et al.* 1994); Budongo is outside the geographical range of the fourth.

Where the same analysis is applied to peninsular Malaysian primate species there is no evidence that any of the four common species are dependent on

Fig. 5.1. A dendrogrammatic representation of primate density interspecific correlation matrices for two tropical forest areas, the Kibale Forest of Uganda (5 study plots: above) and peninsular Malaysia (11 study sties: below). The dashed lines indicate the 5% probability level for *r* (product-moment correlation coefficient). Species closely coevolved with mature forest should branch off at or above the dashed line. Species that benefit from logging should branch off in negative correlation space. (Source: adapted from Skorupa 1986.)

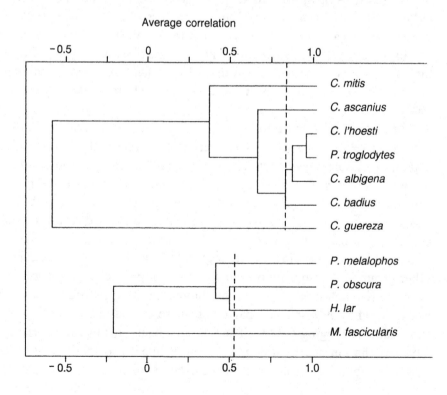

mature forest (Fig. 5.1). Studies of these species at the Tekam Forest Reserve have shown that local variation in density due to local environmental or historical factors can be greater than changes brought about by logging[1-18:51]. Felling operations caused a great deal of mortality among infant primates (Table 5.10), but birth rates quickly returned to normal such that primate densities in logged forests were not reduced. Apart from temporary increases in the density of *Macaca fascicularis*, which tends to invade early successional forest, there are no significant trends in the abundance of the common primates up to 18 years after the felling event (Fig. 5.2).

Two species studied in detail, the lar gibbon *Hylobates lar* and banded leaf monkey *Presbytis melalophos*, both showed decreases in their activity levels following logging, spending less time feeding and travelling and more time resting (Figs 5.3 and 5.4). Gibbons also spent less time in singing, which is energetically expensive. An increase in resting time was necessary to digest the higher proportions of leaf material in their diet. During the first 6 months after completion of logging, the percentage of feeding time spent ingesting leaves increased from 23% to 40% in gibbons, and from 40% to 58% in leaf monkeys).

In unlogged forest gibbons incorporated a high proportion of fruits rich in free sugars into their diet. They were unable to find these fruits immediately following logging and were energetically constrained, reducing day range length by more than 50%. Unlike gibbons, leaf monkeys possess gut adaptations for digesting leaf material. They also reduced day range lengths immediately following logging, but changes in their ranging patterns in logged forest reflected changes in the distribution of food trees rather than energetic constraints. The monkeys were now exploiting common tree species rather than the widely dispersed but highly preferred lipid-rich seeds of trees such as

Table 5.10. *Estimated infant/female ratios among primates at a site in the Tekam Forest Reserve, peninsular Malaysia, monitored between 1978 and 1993*

| Species | Infant/Female ratio | | | | |
|---|---|---|---|---|---|
| | Unlogged | During logging | 1 year after onset of logging | 6.5 years after onset of logging | 12.5 years after onset of logging |
| *Hylobates lar* | 0.50 | 0 | 0.50 | 0.33 | 0.50 |
| *Presbytis melalophos* | 0.41 | 0 | 0.13 | 0.14 | 0.16 |
| *P. obscura* | 0.31 | 0 | 0.09 | 0.20 | 0.09 |
| *Macaca fascicularis* | 0.25 | 0.18 | 0.18 | 0.25 | No data |

*Source*: Grieser Johns & Grieser Johns (1995).

Fig. 5.2. Estimated primate densities up to 18 years post-logging, Tekam Forest Reserve, peninsular Malaysia. The four primates studied are: (a) banded leaf monkey *Presbytis melalophos*; (b) long-tailed macaque *Macaca fascicularis*; (c) dusky leaf monkey *Presbytis obscura*; (d) lar gibbon *Hylobates lar*. Lines link data points from the same study site: closed rectangles, C13C; cross, C5A; star, C1A; open rectangle, C2. Dotted lines indicate the presence of subsistence hunting at this site, which probably affects results. (Source: Grieser Johns & Grieser Johns 1995.)

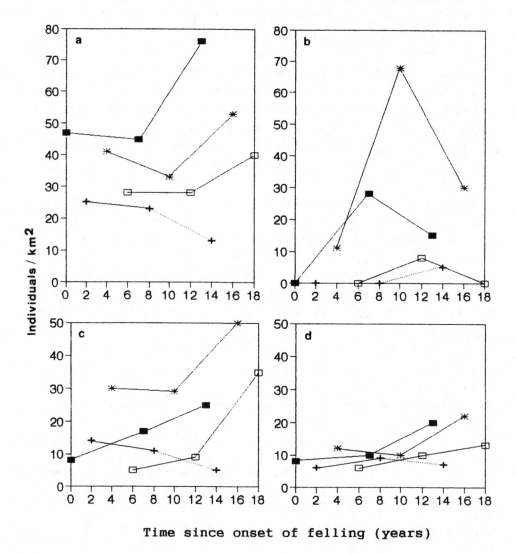

Time since onset of felling (years)

*Parkia* and *Sindora* that were exploited in mature forest. Since food sources were smaller and more evenly dispersed in logged forest, leaf monkey groups split into smaller foraging subunits following logging. This trend towards smaller foraging units continued in the regenerating forest (Grieser Johns & Grieser Johns 1995).

Behavioural changes following logging can lead to quite complex alterations in social organization. In the case of the banded leaf monkeys studied at Tekam Forest Reserve, behavioural adjustments to the changed patterns of food resources in logged forest lead to the abandonment of territoriality and adoption of mutual avoidance as the most efficient means of reducing inter-group competition (Johns 1986b). Facultative territoriality is rare, however, and territorial species typically avoid moving away from their former

Fig. 5.3. Changes in the activity pattern of lar gibbons *Hylobates lar* before and after logging, Tekam Forest Reserve, peninsular Malaysia. Activities (in percentages of total) are (a) resting, (b) feeding, (c) travelling and (d) singing. (Source: Johns 1986b.)

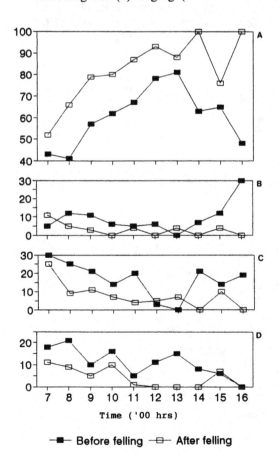

ranges even while logging is occurring. Abandonment of territories occurs only where food resources are critically depleted. Highly territorial animals such as gibbons may remain within their former ranges even following forest clearance or fires which destroy a high proportion of trees (Marsh & Wilson 1981; Beaman *et al.* 1985).

Widespread emigration following logging[1–2:?] has been reported for the territorial indri *Indri indri* of Madagascar (Petter & Peyrièras 1974). This may have been an incorrect conclusion arising from a failure to detect indris in logged forest. The indris may simply have reduced calling rates and adopted cryptic behaviour, as do gibbons. This can result in a species 'reappearing' in the logged forest at some point in time after the logging disturbance has ceased, which is indeed reported in this case.

Fig. 5.4. Changes in the activity pattern of banded leaf monkeys *Presbytis melalophos* before and after logging, Tekam Forest Reserve, peninsular Malaysia. Activities (in percentages of total) are (a) resting, (b) feeding and (c) travelling. (Source: Johns 1986b.)

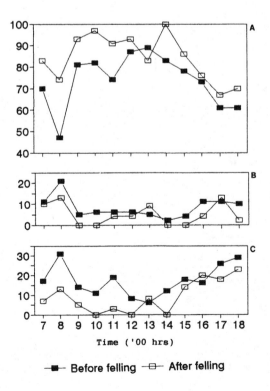

Time ('00 hrs)

—■— Before felling    —□— After felling

### Giant flying squirrels

South-East Asian flying squirrels of the genera *Petaurista* and *Pteromyscus* show anatomical adaptations for folivory despite their fairly small body sizes: 750–2000 g and 200–240 g, respectively (Muul & Lim 1978). In mature forest these squirrels forage primarily for energy-rich soft fruits, but an ability to incorporate leaf material into their diet at times of fruit shortage assists their persistence in logged forest. At the Tekam Forest Reserve, peninsular Malaysia, the proportion of feeding time spent eating leaves by the red giant flying squirrel *Petaurista petaurista* increased from 51% prior to logging to 79% in the 6 months following the completion of logging[0.5:51] (Barrett 1984). In the spotted giant flying squirrel *Petaurista elegans* the change was from 19% to 78%.

Transect surveys suggest that the population densities of *P. petaurista* did not change significantly during the 6 months following logging (from 34 to 39 individuals/km$^2$) but that the population density of *P. elegans* fell (from 22 to 7 individuals/km$^2$). Competition for the remaining preferred energy-rich fruits may have been a factor in the poor persistence of *P. elegans* (Barrett 1984). The densities of both species decreased markedly in older logged forests, despite dietary adaptations that should aid persistence in logged forest. This may reflect an alternative limiting factor, such as the availability of daytime refuges, or perhaps increased predation.

### Ungulates

A number of small tropical forest ungulates feed partly by browsing understorey vegetation (usually early stage regenerating plants in treefall gaps) and partly on fallen fruit. Species such as mousedeer *Tragulus* and muntjac *Muntiacus* appear to be commoner in logged than in mature forest at the Tekam Forest Reserve, peninsular Malaysia. Their abundance increased from 0.5 to 5, and from 1 to 11 individuals/100 km surveyed, respectively, comparing unlogged and logged[5–6:51] forest. Densities tended to decrease again in older logged forests. At the Ulu Segama Forest Reserve in Sabah, however, greater and lesser mousedeer *Tragulus javanicus* and *T. napu* both occurred in reduced densities in logged[5&12:57] forests, while the densities of yellow and red muntjac *Muntiacus atherodes* and *M. muntjac* did not differ significantly between forest types (Heydon 1994).

In Sierra Leone, duikers *Cephalophus* appear well able to persist in logged and secondary forests where they are not hunted (Davies 1987). At the Kibale Forest, Uganda, highest duiker densities were reported in logged[15:? < 35] forest where basal area was 15% less than that of adjacent unlogged forest, but were low in logged[16:? < 50] forest with a 60% reduction in basal area (Nummelin

1990). In Gabon, forest-dwelling duikers did not alter significantly in density as a result of extensive logging operations but there was an invasion of logged[10-15:11] areas by an additional species *Cephalophus sylvicultor* more typically found in savanna habitat (White 1992).

### Rodents

Rodent communities in tropical forests generally contain a moderately large number of species (exceeded only by bats), of which most are terrestrial or semi-terrestrial. Most species feed opportunistically on a variety of plant material and seeds. Some can be important seed predators. In the Kibale Forest of Uganda, changes occurred in the abundance of species in both regenerating[18:? < 35] and degenerating[19:? < 50] logged forest, and colonization by disturbed forest specialists was apparent (Table 5.11). Only one species, a dormouse *Graphiurus murinus*, has been suggested to be adversely affected by intensive logging (Isabirye-Basuta & Kasenene 1987). Overall, there was a

Table 5.11. *Abundances of small rodents in unlogged and logged forest at Kibale Forest, Uganda*

| | No. of marked individuals | | |
|---|---|---|---|
| Species | Unlogged | Regenerating (15% basal area reduction) | Degenerating (60% basal area reduction) |
| *Praomys stella* | 53 | 64 | 65 |
| *P. jacksoni* | 28 | 31 | 39 |
| *Hybomys univittatus* | 7 | 31 | 39 |
| *Graphiurus murinus* | 13 | 22 | 8 |
| *Thamnomys rutilans* | 2 | 5 | 2 |
| *Mus minutoides* | 1 | 1 | 3 |
| *Cricetomys gambianus* | 1 | 1 | — |
| *Lophuromys flavopunctatus* | — | 3 | 3 |
| *Malacomys longipes* | — | 1 | 1 |
| *Dendromys mysticalis* | — | — | 7 |
| *Paraxurus boehmi* | — | — | 9 |
| *Protoxerus stangeri* | — | — | 1 |
| *Funisciurus anerythrus* | — | — | 2 |
| *Rattus rattus* | — | 1 | — |
| Total | 105 | 147 | 167 |
| Trapnights | 6923 | 6969 | 7003 |

*Source*: Adapted from Muganga (1989).

considerable increase in rodent biomass correlated with increased ground vegetation cover in logged areas.

### Insectivores

Foliage insects are a predictable resource in unlogged tropical forest, but less so in logged and secondary forest (see above). Foliage insects are always available within the interior of primary forest, although large fluctuations may occur in open spaces, such as large gaps, and above the canopy. They may become a scarce resource at certain times in logged forest, or such periods may be longer than in primary forest. Periods of low insect abundance may be marked by shifts in the feeding habits of insectivorous species, which may add fruit or increase the proportion of fruit in their diet. Obligate insectivores are not able to compensate in this way.

Because insects are a predictable resource with a regular dispersion in primary forest, insectivores have diversified considerably. This species richness is in proportion to the seasonal stability of the forest vegetation, which changes considerably following logging. Insectivores may be more affected by forest disturbance than many feeding groups since they are not adapted to exploit resources that are both discontinuous in distribution and fluctuating severely over time.

There are three main groups of vertebrate insectivores in tropical forest. These are the Insectivora (comprising groups such as gymnures and shrews), insectivorous bats and insectivorous birds. Least is known concerning the Insectivora. Incidental observations on moonrats *Echinosorex gymnurus* in Malaysia suggests that this species is uncommon everywhere but does occur in low numbers in logged[1-12:51] forest in peninsular Malaysia (Johns 1989a). A more detailed study of tenrecs *Echinops telfairi* and *Tenrec ecaudatus* in western Madagascar demonstrates that neither species is adversely affected by logging[8-9:<50] at fairly low intensities (Ganzhorn *et al.* 1990).

A number of small mammals of other taxa are also predominantly insectivorous. Studies tend to show drops in density in logged forest. For example, the Bornean tarsier *Tarsius bancanus* declined in density by an average of 43% in logged[2-12:57] forests in Sabah (Heydon 1994). There are exceptions, however, such as species able to exploit the large quantities of beetle larvae occupying rotting wood in recently logged forest (e.g. aye-ayes *Daubentonia madagascariensis* of Madagascar and treeshrews *Tupaia* of South-East Asia).

### Understorey birds

Many understorey insectivorous birds are highly specialized, which is necessary in order to feed efficiently on insects that are under strong selection for cryptic colouration or behaviour (Robinson 1969). On average there are 4.3 times as many insectivorous bird species in the tropical forest understorey as in temperate forest understorey (Karr 1976), but population densities of individual tropical forest species are often very low (Orians 1969).

Terrestrial insectivorous species and those foraging in the understorey of mature forest are often vulnerable to microclimatic change, either because they are physiologically ill-adapted to the changed conditions or because their foraging substrata or food supplies are adversely affected. Mechanized logging compacts large soil areas and destroys much understorey vegetation. Although the appearance of pioneers can be rapid, this may not, at least initially, support a full diversity of foliage insects and their predators.

The species richness of understorey insectivores often declines rapidly following logging as birds move away into more favourable habitat. Recolonization can be fairly rapid (e.g. Johns 1989b) but under some conditions it may take considerable lengths of time. Old logged[50:?] dipterocarp forests of southern Vietnam still lack many species of unlogged forest (Kalyakin *et al.* 1994, M. Kalyakin, pers. comm.).

Logging may cause a constriction of niche space, especially for species segregated by foraging height. Among Peruvian antwrens *Myrmotherula*, for example, five species occurred in mature forest but only one in secondary forest; this species *M. brachyura* possessed the broadest foraging niche (Terborgh & Weske 1969). Similarly, changes in the relative abundance of five sympatric trogons *Trogon* occur following logging[1&10:38] in French Guiana, and one, *Trogon rufus*, may be lost (Thiollay 1992). Sympatric congeners generally partition the habitat into narrow niches, and complex groups with many congeners are particularly affected by changes in the diversity of microhabitats.

Five broad foraging groups of understorey insectivores may be identified: terrestrial, foliage-gleaning, bark-gleaning (including woodpeckers), sallying (flycatching) and sallying substrate-gleaning species. Some members of all of these groups are adversely affected by logging. Both the number of species present and their proportional representation within the population sample may decrease following logging at moderate intensities (Table 5.12). The former statistics are minimum figures for species loss since there is occasional substitution of species within the feeding group. Logged forest may be invaded by species not found in mature forest. Precisely which species invade successfully and which species are eliminated is determined largely by microhabitat specializations.

Of the complete avifauna of Sabahan dipterocarp forests, trogons *Harpactes*, woodpeckers Picidae, wren-babblers *Kenopia* and *Napothera*, and flycatchers *Cyornis*, *Ficedula* and others, were identified as prone to reduction in logged[1&8:57] forest by Lambert (1992). All are insectivorous species. These groupings have been revised by Grieser Johns (1996) who compared results from two studies in Sabah and one in peninsular Malaysia to identify consistent trends. Only four common species consistently declined by > 50% following logging[1–18:51–57]: the terrestrial *Kenopia striata* and three flycatchers *Culicicapa ceylonensis*, *Rhipidura perlata* and *Philentoma velatum*. Data are inconclusive for rarer species within the vulnerable groups, however, and the real number of species which consistently decline in logged forest is doubtlessly higher.

Reductions in the abundance of mature forest flycatchers is a common feature of logged forest avifaunas. Understorey flycatchers may be limited partly by changes in the availability of suitable flying insects or by changes in the structure of the habitat such as a shortage of suitable perches. In many cases, however, they may be displaced by flycatcher species more characteristic of edge habitat (e.g. edge-specialist fantail flycatchers in logged dipterocarp forest at Tabin, Sabah, and in mesophyll vine forests on the Australian tablelands: Johns 1989a; Driscoll & Kikkawa 1989). In large gaps, flying insects are also exploited by aerial insectivores such as swifts and swallows which are restricted to foraging above the canopy in mature forest but which invade the foraging volume of understorey flycatchers following logging. These birds are highly mobile and respond quickly to fluctuations in insect abundance (Frith & Frith 1985). Migrating species may also congregate in large gaps in logged forest to take advantage of seasonal peaks in insect abundance. At Ponta da Castanha, Brazil, migrating fork-tailed tyrant-flycatchers *Muscivora tyrannus* and purple martins *Progne subis* congregated in large numbers in logged[11:61] forest and adjacent areas of shifting cultivation. At the Tekam Forest Reserve, peninsular Malaysia, migrating blue-throated bee-eaters *Merops viridis* made up 19% of birds encountered during transect surveys in recently logged[1–2:51] forest.

Neotropical rain forests support a number of highly specialized foliage-gleaning insectivores. Species which feed in association with epiphytes, such as scythbills *Campylorhamphus*, or those specializing in dead-leaf searching, such as some foliage gleaners *Automolus*, may not persist in heavily disturbed habitats. Very large arrays of bark-gleaning woodcreepers Dendrocolaptidae are unique to Neotropical forests. Woodcreeper species segregate partially by foraging height (Brooke 1983) and changes in the relative abundance of species occur following logging, with some species apparently being eliminated. Species substitution following logging also occurs widely among Neotropical woodpeckers Picidae, but this is less common in woodpecker guilds elsewhere. Lambert (1992) suggests Sabahan woodpeckers may be prone to reduction in numbers in logged[1&8:57] forest, although no evidence for this was found by

Table 5.12. *Changes in species richness and abundance of understorey insectivores occurring as a result of selective logging*

| Feeding group | Tree loss (%) | Unlogged | | 1- to 6-year-old logged | | 7- to 12-year-old logged | | 13- to 18-year-old logged | |
|---|---|---|---|---|---|---|---|---|---|
| | | No. species | % total sample | No. species | % total sample | No. species | % total sample | No. species | % total sample |
| **Terrestrial species** | | | | | | | | | |
| **Babblers: Timaliidae** | | | | | | | | | |
| (Tekam F.R.)[a] | 51 | 3 | 0.8 | 1 | 0.05 | 2 | 0.4 | 4 | 0.7 |
| (Ulu Segama F.R.)[b] | 57 | 6 | 4.8 | 3 | 2.0 | 3 | 1.4 | | |
| (Ulu Segama F.R.)[c] | 57 | 5 | 6.2 | | | 3 | 4.0 | | |
| **Foliage-gleaning species** | | | | | | | | | |
| **Babblers: Timaliidae** | | | | | | | | | |
| (Tekam F.R.)[a] | 51 | 16 | 16.3 | 12 | 4.9 | 14 | 10.4 | 15 | 20.4 |
| (Ulu Segama F.R.)[b] | 57 | 14 | 29.5 | 13 | 7.5 | 14 | 13.7 | | |
| (Ulu Segama F.R.)[c] | 57 | 9 | 20.2 | | | 11 | 15.4 | | |
| **Foliage-gleaners: *Automolus*** | | | | | | | | | |
| (Gorupí F.R.)[d] | 10–50 | 3 | 6.0 | 4 | 9.1 | 1 | 0.2 | | |
| (Ponta da Castanha)[e] | 61 | 2 | 2.5 | | | 2 | 0.5 | | |
| (French Guiana)[f] | 38 | 3 | 0.7 | 0 | 3 | 0.1 | | | |
| **Bark-gleaning species** | | | | | | | | | |
| **Woodpeckers: Picidae** | | | | | | | | | |
| (Tekam F.R.)[a] | 51 | 10 | 5.2 | 11 | 2.4 | 9 | 2.6 | 10 | 5.3 |
| (Ulu Segama F.R.)[b] | 57 | 8 | 2.8 | 8 | 1.8 | 8 | 1.8 | | |
| (Ulu Segama F.R.)[c] | 57 | 8 | 2.3 | | | 4 | 0.4 | | |
| (Ponta da Castanha)[e] | 61 | 6 | 1.6 | | | 8 | 3.2 | | |

| | | | | | | | | |
|---|---|---|---|---|---|---|---|---|
| (French Guiana)[f] | 38 | 11 | 1.2 | 14 | 1.9 | 12 | 2.0 | | |
| (Bia)[g] | 15 | 3 | 0.8 | 3 | 0.7 | | | 2 | 0.5 |
| **Woodcreepers: Dendrocolaptidae** | | | | | | | | | |
| (Ponta da Castanha)[e] | 61 | 11 | 7.9 | | | 7 | 6.8 | | |
| (French Guiana)[f] | 38 | 15 | 6.3 | 9 | 2.5 | 12 | 2.3 | | |
| **Sallying species** | | | | | | | | | |
| **Non-migrant flycatchers: Muscicapidae** | | | | | | | | | |
| (Tekam F.R.)[a] | 51 | 9 | 5.8 | 4 | 0.6 | 7 | 1.6 | | |
| (Ulu Segama F.R.)[b] | 57 | 12 | 8.4 | 10 | 3.6 | 8 | 2.8 | 6 | 2.0 |
| (Ulu Segama F.R.)[c] | 57 | 10 | 6.6 | | | 5 | 2.0 | | |
| **Sallying substrate-gleaning species** | | | | | | | | | |
| **Trogons: Trogonidae** | | | | | | | | | |
| (Tekam F.R.)[a] | 51 | 5 | 1.2 | 2 | 0.3 | 2 | 0.2 | | |
| (Ulu Segama F.R.)[b] | 57 | 4 | 1.7 | 2 | 0.5 | 3 | 0.3 | 2 | 0.6 |
| (Ulu Segama F.R.)[c] | 57 | 5 | 1.5 | | | 2 | 0.4 | | |
| (French Guiana)[f] | 38 | 5 | 1.2 | 4 | 2.0 | 4 | 1.7 | | |
| **Puffbirds: *Bucco, Monasa, Malacoptila, Micromonacha*** | | | | | | | | | |
| (Ponta da Castanha)[e] | 61 | 4 | 4.4 | | | 1 | 3.9 | | |
| (French Guiana)[f] | 38 | 3 | 0.7 | 3 | 0.1 | 3 | 0.1 | | |

Data from Tekam F.R. logged forest combines three sites; data from Ulu Segama F.R. combines two sites.

*Sources:* [a]A. Grieser Johns (unpublished data); [b]Grieser Johns (1996); [c]Lambert (1992); [d]Johns (1986a); [e]Johns (1991b); [f]Thiollay (1992); [g]Holbech (1992).

Grieser Johns (1996). There is certainly a high level of bark damage, loss of mossy and lichen cover, and associated reductions in the availability of bark- and crevice-dwelling arthropods. Some species (e.g. *Picus* spp., *Hemicircus con- cretus*) switch to foliage-gleaning in logged forest, however, and there is little evidence of woodpecker species being eliminated.

In Neotropical várzea forests bark-gleaning insectivores are remarkably well represented. The vertical segregation observed in terra firme forest is not very apparent, with all species moving into the canopy of the forest during the high-water season. Bark-gleaning species efficiently exploit the vertical migra- tions of insects caused by changes in water level (Irmler 1979). There is no evidence of a reduction in species-richness, but some reduction in abundance, brought about by logging at a level of 4.6 trees/ha (Johns 1986a). Ten species of Picidae and Dendrocolaptidae were recorded in unlogged and 11 species in logged[ongoing:6] restinga (encounter rates were 93 and 50 individuals/100 km surveyed, respectively). Of other vulnerable insectivore groups, terrestrial and foliage-gleaning insectivores are poorly represented in the várzea avifauna and tyrant flycatchers Tyrannidae show typical decreases in both species richness and abundance following logging even at the low level of tree loss recorded.

### Canopy birds

A number of birds are not bound by microclimatic gradients nor segregated by foraging height but move at random through different canopy levels following the outer surface of the canopy. These species may be observed close to the ground in treefall gaps or in the upper branches of canopy trees. Birds in this group also commonly join mixed flocks, together with understorey bird species. Many species show flexibility of foraging behaviour. For example, the Amazonian canopy woodpecker *Melanerpes cruentatus*, which is common in disturbed habitats, gleans insects from bark or foliage, and adopts flycatching behaviour to feed on swarming termites.

In recently logged forest the foraging volume for canopy insectivorous birds may increase due to the greater surface area of the fragmented canopy. The vertical dispersion of these species changes in logged forest to reflect the shifting location of the bulk of primary productivity, and thus the concentration of foliage insects, in lower canopy levels (Fig. 5.5).

### Commensal insectivores

A reliance on commensalism may be an important factor affecting survival ability. A variety of Neotropical and some African forest birds rely on foraging columns of army ants *Eciton* to flush out suitable insect prey. A loss of ant colonies can rapidly delete bird species from the avifauna of forest remnants

(Lovejoy *et al.* 1984). At Ponta da Castanha, Brazil, the density of ant colonies did not change significantly following logging[11:61] but fewer large insects were flushed from ground vegetation by the ants and the type of insects captured differed.

Professional ant-following birds are typically reduced in density in logged forest (Table 5.13), but this is not necessarily due to an absence of commensals. Species declining significantly may be more affected by microclimatic changes, as are other understorey insectivores. Ant-followers of microclimatically robust taxa, such as the woodcreepers *Dendrocincla fuliginosa* and *D. merula* did not appear to be reduced in density.

A few Neotropical canopy birds associate with foraging monkey groups, attracted by insects flushed out by their passage. Very few species can be regarded as professional monkey-followers, perhaps only the double-toothed kite *Harpagus bidentatus*. At Ponta da Castanha this species was encountered mostly in logged[11:61] forest, due to the concentration in that habitat of foraging groups of squirrel monkeys *Saimiri* and capuchin monkeys *Cebus* (Johns 1991a).

Fig. 5.5. Changes in vertical dispersion of canopy insectivorous birds, Ulu Segama Forest Reserve, Sabah. There is a significant change in vertical dispersion in logged forest (Kolmogorov–Smirnov test: $D_{199,180} = 0.322$, $\chi^2 = 40.27$, $P < 0.001$). The analysis combines wholly insectivorous cuckoos and malcohas, trogons, bee-eaters, dollarbirds, cuckoo-shrikes, minivets, ioras, woodshrikes and drongos. Damage level at the logged forest site was 55%. (Source: A. Grieser Johns unpublished data.)

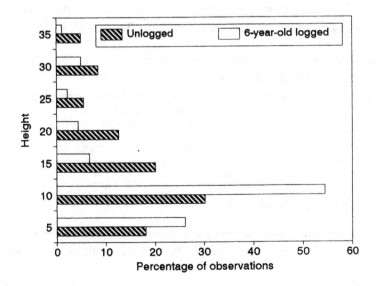

*Bats*

The responses of bats to logging are not well studied. Preliminary data from Queensland suggest that gap connectivity is important in determining the use of logged[1-2:19-22] forest by insectivorous bats (Crome & Richards 1988). Most species are not well adapted to foraging in large artificial gaps caused by logging.

Results from Panama show shifts in the species composition and relative abundance of different groups of insectivorous bats in logged forest that closely parallel changes observed in the avifauna (Johns *et al.* 1985). Small 'fluttering' phyllostomids dropped from 15.3% of mist-net samples in mature forest to 1.7% in forest disturbed by logging and shifting cultivation. They are replaced by species which fly fast and in straight lines such as *Myotis*, *Molossus* and *Molossops*, which normally feed above the canopy in mature forest. The representation of these species in the sample increased from 4% to 15.1% in disturbed forest. A fall in the species-richness of insectivorous bats has also been recorded in logged[7-10:?<50] dipterocarp forests in Sumatra (Danielsen & Heegaard 1995).

Bat populations are highly mobile and can respond quickly to changes in availability of food. Land clearance near the Niah caves in Sarawak caused the number of naked bats *Cheiromeles torquatus* roosting in the caves to rise from practically none to around 350000. This bat is a termite specialist and was exploiting the increase of between 8 and 10 times in the abundance of Macrotermitinae in the cleared forest areas.

### Insectivore-frugivores

The high species richness in tropical forest ecosystems is usually associated with a low level of connectance within the food web. Interspecific

Table 5.13. *Abundances of some professional ant-followers in unlogged and logged forest at Ponta da Castanha, Brazil*

| Species | No. individuals encountered/ 100 km surveyed | |
|---|---|---|
| | Unlogged | Logged |
| *Gymnopithys salvinii* | 3.2 | 2.4 |
| *Percnostola schistaceae* | 2.4 | 0 |
| *Dendrocincla fuliginosa*[a] & *D. merula* | 9.5 | 11.0 |
| *Habia rubica* | 12.2 | 1.5 |

[a]An occasional rather than a professional ant-follower, but difficult to distinguish from *D. merula*.
*Source*: Johns (1991b).

competition in complex ecosystems promotes specialization and tends to eliminate generalists. A predominance of generalists (species which feed from more than one trophic level) is intrinsically destabilizing (Pimm 1980). Results generated from model ecosystems predict that generalists will become abundant following disturbance but that this will be a temporary phenomenon, assuming that regeneration is successful (Pimm & Lawton 1978). If generalists remain abundant, it can be assumed that natural regeneration processes have been arrested. Monitoring the abundance of generalists can therefore have important implications for forest management strategies.

Generalists of tropical rain forests are mostly insectivore-frugivores (there are a few frugivore-predators, such as certain civets Viverridae: see below). Insectivore-frugivores occur in mature tropical forest but usually in low numbers, and often associated with early successional patches and riverine or edge habitat. Following disturbance to mature forest, some of these species, particularly small-bodied species, may show rapid increases in abundance through immigration from surrounding habitats.

True insectivore-frugivores regularly consume approximately equal amounts of food from different trophic levels. Many more species feed preferentially from one level but have the ability to shift dietary composition if necessary. A large number of rain forest frugivores may occasionally eat insects, although their overall foraging and anatomical adaptations are for frugivory. For example, white uakaris *Cacajao c. calvus* in western Amazonia feed extensively on caterpillars of a noctuid moth at times of fruit scarcity, caterpillars being more digestible than foliage (Ayres 1986b). Conversely, prinias *Prinia* and tailorbirds *Orthotomus* of South-East Asia are essentially foliage-gleaning insectivores, but feed on fruit in recently logged forest at times when insect abundance is low.

Not all insectivore-frugivores increase in abundance in logged forest. Large terrestrial phasianids are characteristically less abundant in logged forest due to changed microclimatic conditions and a loss of the litter insects important in their diet (Johns 1989b, Holbech 1992, Thiollay 1992). Species may be eliminated from logged forest areas where extensive hunting by local human populations occurs. In non-hunted forest, recovery of populations usually takes place once a tree canopy is re-established.

### Primates

Studies of species-rich primate communities have attempted to explain sympatry through various mechanisms. In West Africa, individual species within mixed cercopithecine groups diverge in diet and foraging behaviour only at times of critical food shortage, but segregation is then apparent by feeding niches. Divergent feeding ecology and use of different substrata is evident among sympatric Asian forest primates. In Neotropical primates, divergence through microhabitat specialization is more evident: this was demonstrated by six out of eight small-bodied species in Amazonian Peru (Terborgh 1983). Small-bodied species (*Saguinus, Callithrix, Saimiri* and *Cebus*) are able to specialize in exploiting early successional and edge habitats through adaptive dietary habits: they are opportunistic frugivore-insectivores (Johns & Skorupa 1987). Metabolic demands faced by large species are too great to enable specialization in patchy subhabitats. Individual patches would be too small and too much energy would be expended in travelling between them.

At Ponta da Castanha, Brazil, *Saguinus, Saimiri* and *Cebus* occurred at higher densities in logged[11:61] forest (Table 5.14), although most of the groups also ranged into unlogged forest. Foraging in logged forest was typically dispersed rather than cohesive and directed towards arthropod prey and small fruit sources rather than large fruiting canopy trees. In peninsular Malaysia, the prosimian slow loris *Nycticebus coucang* foraged primarily for insects in recently logged[1:51] forest (93% of foraging time, compared with 36% of foraging time prior to logging: Barrett 1984), and its population densities in logged areas, although patchy, do not appear to decrease.

Various Asian macaques *Macaca*, have been reported to develop diverse foraging behaviour in logged or disturbed forest. Following a prolonged drought in Sri Lanka, a group of toque macaques *Macaca sinica* began exploiting animal and other food items from a rubbish tip within its home range (Dittus 1977). In logged[1:?>50] mangrove forest in peninsular Malaysia, long-tailed macaques *Macaca fascicularis* foraged mainly on the ground, ate an unusually high proportion of leaf material (33% of feeding observations) and also scavenged from a rubbish tip (48% of feeding observations) (Lim & Sasekumar 1979). While such commensalism with human populations is highly

Table 5.14. *Population densities of insectivorous-frugivorous primates in unlogged and logged forest at Ponta da Castanha, Brazil*

| Species | Individuals/km$^2$ | |
| --- | --- | --- |
| | Unlogged | 11-year-old logged |
| *Saguinus mystax* | 78 | 88 |
| *Saimiri* sp. | 32 | 81 |
| *Cebus apella* | 11.5 | 32 |
| *Cebus albifrons* | 14 | 31 |

Tree loss rate during logging was 61%.
*Source*: Johns (1991a).

unusual in tropical forest species, it represents an extreme of dietary adaptation which will allow such species to persist in even intensively logged forests.

### Transient understorey birds

The avifauna of early regenerating forests often shows over-dominance by a few species of insectivore-frugivores rarely seen in mature forest. In the Neotropics, secondary growth is colonized by tanagers Thraupidae and finches Fringillidae, the latter being gramnivores. In Africa and Asia, over-dominant species are typically greenbuls and bulbuls Pycnonotidae.

These birds are transient species, congregating in early successional habitats. They are essentially light-seeking, associated with rapidly growing pioneer species providing berry-like fruit. Since treefall gaps are continually forming, they remain in regenerating forest as it matures but their numbers will reflect the density of early successional patches or regeneration stages. Potentially successful regeneration at two Malaysian dipterocarp forest sites results in the densities of most species becoming less in older logged forests (Table 5.15). In this example, a succession of bird species is also apparent, with the yellow-vented bulbul *Pycnonotus goiavier* reaching a much earlier peak than other species, particularly the yellow-breasted flowerpecker *Prionochilus maculatus*.

In the Kibale Forest of Uganda, degenerating logged[23:? < 50] forest retains an over-dominance of yellow-whiskered greenbuls *Andropadus latirostris* which made up 26% of understorey birds mist-netted (Dranzoa 1995). This probably reflects the poor forest regeneration. However, this greenbul is a lekking species and the historical location of leks has been shown to have a significant effect on abundances in areas of different logging histories at the Budongo Forest, Uganda (A. Plumptre, pers. comm.).

Fluctuations in abundance of these understorey birds contrast with those of canopy insectivore-frugivores, such as certain barbets Capitonidae and orioles *Oriolus*, which also exploit the fruit of pioneers but whose numbers tend to remain constant over mature and logged forest.

### Predators

Predators, located at the top of the food chain, might be predicted to be highly susceptible to disturbances affecting the food web. However, most species exploit a variety of prey species opportunistically and are mobile animals with large home ranges: characteristics which are likely to improve their adaptability to changes caused by logging. They tend to congregate where food is most abundant or most easily captured. Thus jaguars *Felis onca* have tended to congregate around newly established cattle ranches in Amazonia where they feed on calves and foals. Philippine eagles *Pithecophaga jefferyi* are most abundant in mosaics of primary and logged forest, and capture most of their prey, particularly colugos *Cynocephalus variegatus*, in logged forest where they are more easily seen.

The construction of roads and landing areas, and the fragmentation of the forest canopy by logging, increase the visibility of many prey species. Small animals are forced to cross open areas more frequently in logged forest. In peninsular Malaysia, patrolling or scanning of newly constructed logging roads was observed in many raptors, including collared scops owls *Otus bakkamoena*, which capture large beetles, and hawk-eagles *Spizaetus* and serpent eagles *Spilornis*, which catch mostly reptiles. During logging and forest clearance in the catchment area of the Tucuruí dam in Amazonia, a high density of harpy eagles *Harpia harpyja* was recorded in the project area (Johns

Table 5.15. *Abundances of selected insectivorous-frugivorous birds in unlogged and logged dipterocarp forest*

| Area and species | | Individuals encountered/100 km surveyed | | | |
|---|---|---|---|---|---|
| | Unlogged | 0- to 6-month-old logged | 1- to 6-year-old logged | 7- to 12-year-old logged | 13- to 18-year-old logged |
| Tekam Forest Reserve[a] | | | | | |
| *Pycnonotus atriceps* | 1 | 1 | 53 | 27 | 35 |
| *P. squamatus* | 8 | 6 | 173 | 9 | 9 |
| *P. goiavier* | 0 | 2 | 136 | 73 | 0 |
| *P. simplex* | 13 | 16 | 640 | 64 | 109 |
| *P. brunneus* | 18 | 20 | 128 | 403 | 249 |
| *Hypsipetes criniger* | 21 | 3 | 60 | 284 | 183 |
| *Prionochilus maculatus* | 0 | 0 | 64 | 22 | |
| Ulu Segama Forest Reserve[b] | | | | | |
| *Pycnonotus atriceps* | 13 | | 30 | 142 | |
| *P. eutilotus* | 1 | | 23 | 23 | |
| *P. simplex* | 4 | | 52 | 67 | |
| *P. brunneus* | 18 | | 69 | 61 | |
| *P. erythrophthalmus* | 4 | | 18 | 112 | |
| *Prionochilus maculatus* | 10 | | 15 | 30 | |

Except for results from 0- to 6-month-old logged forest, logged forest densities are significantly higher than unlogged (Wilcoxon–Mann–Whitney tests, $P < 0.05$).
[a]Figures from logged forest are combined data from three sites; tree loss rate was 51%.
[b]All figures are combined data from two sites; tree loss rate was 57%.
*Source*: A. Grieser Johns (unpublished data).

1986a). The eagles captured animals moving between remaining forest patches. The abundance of raptors tends to decrease as forest undergrowth regrows and covers road surfaces. In peninsular Malaysia and Sabah the abundance of raptors had reverted to that in unlogged forest by about 6–7 years after the logging event

In the Kibale Forest of Uganda, the predation of monkeys by both crowned eagles *Stephanoaetus coronatus* and chimpanzees *Pan troglodytes* increased in logged forest (Skorupa 1988). In the broken canopy monkeys were more vulnerable to attack by the eagles, and when they were forced to move along or close to the ground to reach food trees they were more vulnerable to attack by the chimpanzees. The red colobus *Colobus badius* developed behavioural adaptations to reduce the risk of predation by chimpanzees, forming highly aggressive male coalitions in logged forest which would attack even human observers.

Congregation of browsing mammals such as deer and tapirs, which feed on the ground vegetation of recently logged forest, caused associated rises in densities of jaguars *Felis onca* in Peru (M. Dourojeanni, pers. comm.) and of tigers *F. tigris* in peninsular Malaysia (Johns 1983). On a smaller scale, the much increased densities of ranid frogs and rats (particularly *Rattus tiomanicus*) in recently logged forest in peninsular Malaysia increased the occurrence of small cats *Felis bengalensis* and *F. marmorata* by four times (Johns 1983). Snakes such as cobras in Asia and boas in the Neotropics can also be attracted into recently logged forest.

The only group of predators that are consistently adversely affected by logging are piscivores. There are relatively few members of this group, but piscivorous kingfishers Alcedinidae are usually absent from recently logged forest due to deterioration of water quality and reduction in the abundance or ease of capture of prey items.

### Civets

Up to eight species of civet Viverridae may occur sympatrically in dipterocarp forests of South-East Asia. Most species feed opportunistically on a variety of small animals, although the otter civet *Cynogale bennettii* is adapted to locating and feeding on aquatic crustaceans. Diets of some civets may also include substantial quantities of fruit, and they might more accurately be categorized as frugivore-predators.

In Sabah, Borneo, Heydon & Bulloh (1996) recorded a marked decrease in civet density from 31.5 individuals/km$^2$ in unlogged to 6.4 individuals/km$^2$ in logged[12:57] forests (Table 5.16). Predominantly carnivorous species were reduced in density to a greater extent than palm-civets, which incorporate larger quantities of fruit into their diet. Few data are available for the otter civet, but it is likely that its requirement for clear, fast-flowing shallow streams would make it particularly susceptible to logging disturbance.

In peninsular Malaysia, Johns (1983) recorded civets as being extremely scarce in unlogged forest (only the binturong *Arctictis binturong* was recorded in 16 months of study). Immediately following the onset of logging[0–0.5:51], however, several additional species of civet moved into the area. Densities of eight individuals/km$^2$ were recorded by 6 years post-logging and 17 individuals/km$^2$ by 12 years post-logging (six species were now recorded). Why the results of this study should be fundamentally different to those from Sabah are unclear. Volumes of timber extracted were twice as high in Sabah, but overall damage levels not dissimilar between the two sites (51% and 57% tree loss in peninsular Malaysia and Sabah, respectively).

Table 5.16. *Sighting frequencies and estimated total densities of civets in unlogged and logged dipterocarp forest (Ulu Segama, Sabah)*

| Subfamily and species | Animal material in diet (%) | No. sightings/100 km surveyed | | |
| --- | --- | --- | --- | --- |
| | | Unlogged | 6-year-old logged | 1- to 12-year-old (six sites combined) |
| Paradoxurinae (palm civets) | | | | |
| *Paradoxurus hermaphroditus* | | 9.2 | 0 | 3.1 |
| *Arctogalidia trivirgata* | <60 | 4.1 | 8.8 | p |
| *Arctictis binturong* | <60 | 1.0 | 0 | 2.4 |
| *Paguma larvata* | 60–95 | p | 0 | p |
| Hemigalinae (banded palm-civets) | | | | |
| *Hemigalus derbyanus* | 100 | 12.3 | 4.4 | 2.2 |
| *Cynogale bennettii* | 100 | 1.0 | 0 | p |
| Viverrinae (true civets, linsangs) | | | | |
| *Viverra tangalunga* | 88 | 20.5 | 8.8 | 6.1 |
| *Prionodon linsang* | >95 | 3.1 | 0 | p |
| Total distance surveyed (km) | | 97.5 | 23.0 | 72.3 |
| Total civet density calculated by Heydon & Bulloh (n/km$^2$)[a] | | 31.5 | — | 6.4 |

Sighting rates are those of Heydon & Bulloh (1996) and A. Grieser Johns (in Heydon & Bulloh 1996) combined.
[a]Unlogged and logged densities are significantly different (Wilcoxon–Mann–Whitney test, $P < 0.05$).
*Source*: Adapted from Heydon & Bulloh (1996).

### Decomposers

A combination of equable temperatures and humidity in the litter of mature forests has led to a greater proportion of decomposition being performed by fungi than is typical of temperate forests. Some species of litter invertebrates, such as diplopods (millipedes) are thus fungivorous in the tropics rather than themselves decomposers. Most decomposition in tropical forest is carried out by fungi in symbiosis with isopterans (termites) and is highly efficient. For example, 100% of leaf litter has been reported to be decomposed annually in Malaysia and 1.3%/day in Ghana.

Most termites are specialized in their feeding habits and their nests are often small and vulnerable to damage during logging. Rotten wood and soil feeders of the Rhinotermitidae, Termitinae and Nasutotermitinae rely on the high levels of fungal and microbial decay typical of rain forest litter to pre-rot plant material to a palatable level. The Macrotermitinae maintain a fungus *Termitomyces* inside their nests. The fungus increases the efficiency of termite feeding by completing nutrient release within partly digested faecal matter, which is subsequently re-ingested by the hosts.

Increased ground temperatures in logged forest may adversely affect the eggs and larvae of soil termites building simple nests. Mound-builders are at an advantage as the complex nest structure buffers changes in ground and air temperatures. Drying out of the soil also acts to slow down fungal and bacterial decay, which can affect species that do not maintain *Termitomyces* within temperature- and humidity-regulating mounds.

The species-richness of termites is much reduced in logged forests (e.g. Sarawak[1: > 50]: Collins 1980; Cameroon[0.5–30:?–100]: Eggleton *et al.* 1995), due primarily to a loss of soil feeders. The numerical abundance of mound-building Macrotermitinae tends to increase considerably. The removal of fresh litter by these species in logged forest reduces the amount of humus and alters the organic content of the soil. In intensively logged areas this is potentially a significant factor affecting nutrient cycling.

Amongst other decomposers, annelids (earthworms) tend to have a patchy distribution in logged forest, being absent from compacted soils and those receiving direct sunlight. In general terms, soil organisms persist better in the litter remaining underneath residual trees and under logging debris than in damaged areas.

In mangrove forests, a high proportion of fresh leaf litter is either eaten by crabs or buried by them, forming peat. Between 40 and 90% disappears after 20 days (Whitten *et al.* 1984). In mangrove forests subjected to logging the abundance of these crabs can be reduced and decomposition conducted to a larger extent by microorganisms. Decomposition then takes much longer, perhaps 4–6 months, and patterns of nutrient recycling are affected accordingly.

### Summary

Changes in the abundance of individual species following logging can result from changes in food availability, microclimatic or other environmental conditions, or occasionally from changes in competitive relationships. Logging rarely acts to eliminate species, especially where damage levels are low, but considerable changes may occur in the relative abundance of species.

If logging acts to decrease the resources available to a species that species should be reduced in abundance in logged forest. On rare occasions, the abundance of particular species may be correlated with that of particular resources and their response becomes predictable. The abundance of certain primates may co-vary with the abundance of particular fruit trees, and if these are eliminated by logging then the primate is also eliminated, if it is unable to switch to other food sources. Terrestrial and understorey insectivores are often adversely affected by logging which reduces available insect resources through a combination of microclimatic and physical environmental effects, and through increasing competition with canopy and above-canopy species of birds and bats.

Conversely, if logging acts to increase the resources available to a species, that species should increase in abundance. Its rate of increase will be influenced by its mobility or breeding level. An increase in the abundance of pioneer trees in regenerating forest is reflected in increases in associated foliage insects and in their predators. Flowing exudate from trees damaged during logging attracts exudate feeders. The abundance of terrestrial herbaceous vegetation in early regeneration stages attracts browsing mammals. Increases in species associated with early regeneration stages or edge vegetation should be a temporary phenomenon. If they persist, it is a demonstration of poor forest regeneration.

The species most likely to maintain or increase their numbers in logged forest are those specialized to secondary vegetation or those with highly flexible diets. Frugivores and nectarivores are wide-ranging and adapted to exploit resources that are widely distributed and highly seasonal. However, they may be prone to reduction or elimination of food resources below a level at which they can be exploited efficiently. Frugivore/folivores, on the other hand, are likely to be adaptable to such changes, although they may alter social behaviour, foraging group sizes and ranging patterns to suit the new distribution of resources. Similarly, frugivore/insectivores are in most cases well able to shift diets to adjust to changed conditions in logged forests.

# 6

## Responses of species assemblages

### Introduction

Studies of the ecology of individual species can help to identify changes in feeding, ranging or other behaviour following logging, and how this may affect their population densities in logged forest. This may be important information for devising conservation strategies directed towards individual species (e.g. orang-utans in Sabah: Payne 1988). However, it is not possible to generalize from a study of a single species to management of biodiversity in a wider context. For management of biodiversity resources as a whole it is more appropriate to examine species assemblages with a view to obtaining more wide-ranging information that can be incorporated into management planning.

The extent of change within species assemblages will vary in broad terms according to the levels of disturbance. Thus a group of central African species subjected to very light levels of logging is likely to change less than an equivalent group of South-East Asian species subjected to very high logging intensities. However, there are also geographical differences in the structure of rain forest communities (Karr 1980, 1990; Bourlière 1989), reflecting different regional evolutionary processes. These differences can influence the extent of immediate change and the time taken for the community to return to its pre-logging state. For example, island forests within the cyclone belts of the Philippines and Melanesia are naturally adapted to recover quickly from extremely high levels of damage. Although the need for intervention to improve biodiversity conservation is likely to be less urgent under conditions of light logging, there is no linear relationship between logging intensity and the extent of change in species assemblages.

There are three approaches to studies of species assemblages. The first is to survey the species present, estimate their densities, and then use a basic knowledge of the ecology of species to examine trends within species groups. This enables a rapid assessment of groups that may respond poorly to defined management practices and thus require some form of conservation action. This might take the form of increasing the extent of totally protected forest estate, defining corridors connecting remnant forest patches, or emphasizing the well-known need to reduce damage levels during felling and transport operations (Palmer & Synnott 1992).

An extension of this approach is to attempt to determine causal relationships between changes in biodiversity and vegetational/environmental parameters. Although the impacts of logging operations have often been described, they have rarely been analysed and the precise remedial measures determined. This type of information, which is of primary use in developing improved forest management, is currently very rare and can generally be obtained only from long-term studies of species assemblages (see Chapter 7).

The third approach is to attempt to define features of the species assemblages, or to identify subcomponents of the assemblage, that can be used to impart information concerning the management practices. For example, changes in species diversity in logged forest, the extent of overlap between unlogged and logged forest, or the behaviour of defined 'indicator species', may be used to predict whether or not animal populations can be sustained under the applied forest management system. If adverse responses are recorded, however, this approach does not supply information as to what specific action may be required.

This chapter reviews existing data on the responses of species assemblages to logging and examines the efficiency of various methods of describing these responses.

### Changes within species assemblages
*Litter invertebrates*

Invertebrate communities in tropical forests have been studied to a limited extent, reflecting problems of field taxonomy. Conclusions based on identifications at the level of taxonomic grouping only are clearly of limited value in assessing changes. Data from Sabah suggest that some changes in the relative abundance of soil and litter invertebrates occur following logging (Table 6.1) but that invertebrate abundance in logged forest was highly variable according to litter density and humidity. In the more detailed study of litter invertebrates by Burghouts *et al.* (1992), only four invertebrate groups showed significant changes in abundance following logging[11:57] (mites, pseudoscorpions, scorpions and termites, all of which decreased). Groups of predatory species tend to be reduced in abundance, although for many, such as spiders, the decrease is not significant. The abundance of termites in the litter layer may increase, but their abundance in the soil increases considerably.

In general, soil and litter invertebrates respond to very local microclimatic changes caused by logging rather than to changes in the structure or

Table 6.1. *Abundance of soil and litter invertebrates in unlogged and logged dipterocarp forest at Ulu Segama Forest Reserve, Sabah*

| | No. individuals/m$^2$ | | | |
| --- | --- | --- | --- | --- |
| | Soil and litter[a] | | Litter[b] | |
| Species group | Unlogged | 5-year-old logged (47% tree loss) | Unlogged | 11-year-old logged (60% tree loss) |
| Crustacea | | | | |
|   Isopoda (woodlice) | 28 | 72 | 71 | 72 |
| Arachnida | 30 | 25 | | |
|   Acarina (mites) | | | 54 | 29 |
|   Araneae (spiders) | | | 74 | 71 |
|   Phalangida (harvestmen) | | | 6 | 4 |
|   Pseudoscorpionida | | | | |
|     (pseudoscorpions) | | | 56 | 18 |
|   Scorpionidae (scorpions) | | | 1 | 0 |
| Myriapoda | | | | |
|   Chilopoda (centipedes) | 9 | 19 | 15 | 13 |
|   Symphyla | 40 | 16 | | |
|   Diplopoda (millipedes) | 12 | 35 | 27 | 28 |
| Insecta | | | | |
|   Blattodea (cockroaches) | 9 | 19 | 6 | 10 |
|   Coleoptera (beetles) | 47 | 9 | 61 | 80 |
|   Collembola (springtails) | | | 252 | 184 |
|   Formicidae (ants) | 53 | 78 | 465 | 364 |
|   Hemiptera (bugs) | | | 19 | 17 |
|   Isoptera (termites) | 1215 | 6278 | 60 | 24 |
|   Orthoptera (crickets) | | | 5 | 6 |
|   Dermaptera (earwigs) | 0 | 3 | | |
| Annelida (earthworms) | 12 | 2 | 14 | 9 |
| Mollusca (snails) | 0 | 3 | | |

[a]Samples include litter and 30-cm depth of soil beneath the litter.
[b]Samples include litter only.
*Sources*: J. Anderson in Royal Society (1987); Burghouts *et al.* (1992).

composition of the forest as a whole. Their distribution through logged forest is therefore very heterogeneous.

### Birds

While tropical forest birds have been relatively well studied, detailed information on community structure is rare. There are three reasons for this. First, a very large number of species co-occur in tropical forests. More than 500 species occur within a 97-ha study plot of floodplain forest at Cocha Cashu, Peru (Terborgh *et al.* 1990); 364 species have been recorded in a 200-ha plot of lowland forest at M'Passa, Gabon (Brosset 1990). Second, many species are rare. The majority of forest birds have patchy distributions correlating with patchiness of microhabitats. At Cocha Cashu only 84 species (26% of the total) occurred at densities $\geq 1$ pair/km$^2$. Since field studies rarely cover large areas, study plots are unlikely to cover the full range of microhabitats and thus the full avifauna. Third, many species have cryptic behaviour patterns. A variety of census techniques has been developed to attempt to counter this but none are able to sample the full range of species present (visual counts overlook cryptic species, mist-netting does not sample canopy species, plotting vocalizations overlooks non-vocal species, etc.).

Most data comparing the communities of unlogged and nearby logged forests result from short-term studies in rather small study plots. Studies also tend to compare different sites with different logging histories and are thus biased by patchiness in the original (unlogged forest) avifauna (see below). The results of these studies are non-standardized and generally record rather low numbers of bird species. They are useful, however, in assessing changes in the relative abundance of species or individuals within feeding guilds and thus in interpreting the response of the community as a whole.

Results from selected studies in the three main tropical forest regions (Table 6.2) are chiefly remarkable in the lack of change in distribution of the recorded species among feeding guilds in unlogged and logged forests. There are no overall significant differences between logged and unlogged forests and no consistent changes in guild membership.

The numbers of species recorded from small study areas often decline in logged forest, although there are exceptions. Results from a long-term monitoring programme in peninsular Malaysia (Table 6.3) clearly indicate a drop in species-richness within the population sample. Species-richness appeared to drop in all guilds and had not recovered by 12–18 years post-logging.

Drops in species richness recorded from population samples need not imply that species are becoming deleted from the avifauna, however. In logged[1&10:38] forest in French Guiana, Thiollay (1992) records 39% of the avifauna as being reduced in numbers by $\geq 50\%$; 28% and 31% of bird species were absent from forests logged 1 and 10 years previously. At the Tekam Forest Reserve, 22% of birds were reduced by $\geq 50\%$ after logging[1–13:51], and up to 34% of species were absent from population samples in logged areas (Table 6.3). However, only 2% of the formerly resident species fail to occur in any logged[1–18:51] forest site at Tekam. Studies in various Australian tropical and subtropical forests have found no bird species to be eliminated by logging, although small bird density declined by as much as 50% (e.g. Western Australia[1:50]: Abbott & Heurck 1985; Queensland[1:19–22]: Crome 1991; Tasmania[1–3:30–50;17:100]: Taylor & Haseler 1995). Recorded losses of species may in reality be a failure to record species which become rarer or range more widely in logged than in unlogged forest. These species would be recorded only by extensive or very long-term studies.

While there is little evidence of change at the level of species, there are considerable changes in the relative abundance of species between guilds (Table 6.4). As predicted from ecological characteristics of regenerating forest, there are increases in the abundance of generalists (arboreal insectivore-frugivores and insectivore-nectarivores) and corresponding decreases in bark-gleaning and particularly sallying insectivores. The abundance of terrestrial insectivores appears to recover in older logged forests while that of raptors peaks in recently logged forest but rapidly falls off again.

Dividing birds into feeding guilds is somewhat arbitrary, however, since seasonal shifts in diet and some level of flexibility to changes in the relative abundances of food types may be common. Body size may also be a contributing factor. In French Guiana 50% and 88% of large birds ($\geq 170\,g$ body weight) were absent from logged[1&10:38] forest, compared with 23% and 30% of small birds ($< 170\,g$ body weight). Large body size tends to correlate with specialization in food type and feeding behaviour (Cope's Law) but large-bodied species are spread over a number of feeding guilds.

### Mammals

For similar reasons to bird communities, mammal communities can be very species-rich in tropical forests. Most species richness occurs among small-bodied mammals, however. In the Neotropics and South-East Asia, diversity is particularly pronounced among bats: 65 species of bats occur sympatrically in Costa Rican lowland forest and 45 species occur sympatrically in lowland dipterocarp forests of Sabah. In Africa, the relative diversity of primates and murid rodents is particularly high (up to 14 species of each occur sympatrically in the lowland forests of Gabon). Of large-bodied mammals (i.e.

Table 6.2. *Some examples of the division of bird species between feeding guilds in unlogged and logged forests*

| Feeding guild | Bia Forest, Ghana | | | Ulu Segama, Sabah | | | Ponta da Castanha, Brazil | |
|---|---|---|---|---|---|---|---|---|
| | Unlogged[a] | 2-year-old logged (15% tree loss)[a] | 14-year-old logged (15% tree loss)[a] | Unlogged[a] | 6-year-old logged (57% tree loss)[a] | 12-year-old logged (58% tree loss)[a] | Unlogged | 11-year-old logged (60% tree loss) |
| Frugivores | | | | | | | | |
| Terrestrial | 0 | 0 | 0 | 2 | 2 | 2 | 4 | 3 |
| Arboreal | 12 | 12 | 13 | 9 | 8 | 9 | 21 | 12 |
| Faunivore-frugivores | 7 | 7 | 8 | 7 | 7 | 7 | 0 | 0 |
| Insectivore-frugivores | | | | | | | | |
| Terrestrial | 5 | 4 | 2 | 3 | 3 | 1 | 8 | 6 |
| Arboreal | 16 | 16 | 15 | 25 | 24 | 28 | 19 | 14 |
| Insectivore-nectarivores | 5 | 6 | 5 | 10 | 11 | 12 | 8 | 6 |
| Insectivores | | | | | | | | |
| Terrestrial | 4 | 3 | 4 | 14 | 7[c] | 7 | 11 | 10 |
| Bark-gleaning | 3 | 3 | 2 | 10 | 9 | 9 | 18 | 15 |
| Foliage-gleaning | 47 | 43 | 47 | 44 | 47 | 39 | 25 | 16 |
| Sallying | 8 | 11 | 10 | 19 | 15 | 14 | 17 | 10 |

| Carnivores | | | | | | | |
|---|---|---|---|---|---|---|---|
| Raptors | 4 | 3 | 8 | 5 | 10 | 6 | 9 | 7 |
| Piscivores | 0 | 0 | 0 | 7 | 3 | 5 | 0 | 0 |
| No. species | 111 | 108 | 114 | 155 | 146 | 139 | 140 | 99 |
| No. individuals | 3017 | 3222 | 3383 | 2028 | 1571 | 1595 | 2183 | 647 |
| No. species in sample of 2000 birds | 104[b] | 105[b] | 95[b] | 153 | 150[b] | 150[b] | 140 | 144[b] |
| Comparison with unlogged (χ² value) | | 1.23 | 3.02 | | 6.55 | 3.21 | | 1.31 |
| P | | >0.05 | ≥0.05 | | >0.05 | >0.05 | | >0.05 |

[a]Results from two sites combined.
[b]From logarithmic model of cumulative species abundance.
[c]χ² subcell value >1.0.
Sources: Holbech (1992); A. Grieser Johns (unpublished data).

Table 6.3. *Changes in the distribution of species between guilds over time at Tekam Forest Reserve, peninsular Malaysia (51% tree loss)*

| Feeding guild | | No. species observed | | | |
|---|---|---|---|---|---|
| | Unlogged | 1–6 months after logging | 1–6 years after logging[a] | 7–12 years after logging[a] | 13–18 years after logging[a] |
| Frugivores | | | | | |
| Terrestrial | 2 | 1 | 1 | 1 | 1 |
| Arboreal | 16 | 10 | 9 | 15 | 13 |
| Faunivore-frugivores | 6 | 5 | 7 | 7 | 6 |
| Insectivore-frugivores | | | | | |
| Terrestrial | 5 | 1 | 2 | 0 | 1 |
| Arboreal | 28 | 23 | 30 | 23 | 24 |
| Insectivore-nectarivores | 10 | 8 | 9 | 8 | 8 |
| Insectivores | | | | | |
| Terrestrial | 16 | 6 | 4[c] | 7 | 6 |
| Bark-gleaning | 11 | 7 | 11 | 9 | 11 |
| Foliage-gleaning | 55 | 40 | 41 | 43 | 36 |
| Sallying | 20 | 17 | 9 | 14 | 12 |
| Carnivores | | | | | |
| Raptors | 15 | 9 | 11 | 8 | 3 |
| Piscivores | 1 | 0 | 1 | 0 | 1 |
| No. species observed | 184 | 128 | 135 | 135 | 122 |
| No. individuals | 1804 | 1723 | 2263 | 2748 | 1312 |
| No. species in sample of 2000 birds | 189[b] | 135[b] | 135 | 125 | 137[b] |
| Comparison with unlogged ($\chi^2$ value) | | 2.66 | 8.73 | 3.57 | 7.07 |
| $P$ | | >0.05 | >0.05 | >0.05 | >0.05 |

$\chi^2$ tests are performed on original data (frugivores combined, insectivore-frugivores combined, carnivores combined). Table excludes aerial insectivores (swifts and swallows) and migrants.

[a] Results from three sites combined.
[b] From logarithmic model of cumulative species abundance.
[c] $\chi^2$ subcell value >1.0.
*Sources*: Johns (1986c); A. Grieser Johns (unpublished data).

Table 6.4. *A comparison of feeding guild membership within samples of individual birds at Tekam Forest Reserve, peninsular Malaysia (51% tree loss)*

| Feeding guild | | Percantage of sample | | |
|---|---|---|---|---|
| | Unlogged | 1–6 months after logging | 6–7 years after logging | 12–13 years after logging |
| Frugivores | | | | |
| Terrestrial | 0 | 0.4 | 0 | 0 |
| Arboreal | 10.6 | 10.6 | 14.9 | 11.9 |
| Faunivore-frugivores | 11.5 | 10.3 | 5.8 | 6.6 |
| Insectivore-frugivores | | | | |
| Terrestrial | 0.6 | 0 | 0 | 0.2 |
| Arboreal | 15.1 | 8.8 | 39.1 | 31.0 |
| Insectivore-nectarivores | 3.9 | 2.0 | 4.5 | 8.9 |
| Insectivores | | | | |
| Terrestrial | 3.6 | 1.4 | 0.1 | 2.4 |
| Bark-gleaning | 5.3 | 3.4 | 3.0 | 3.8 |
| Foliage-gleaning | 31.1 | 34.5 | 28.5 | 28.4 |
| Sallying | 16.2 | 24.1 | 2.0 | 5.2 |
| Carnivores | | | | |
| Raptors | 2.0 | 4.5 | 2.0 | 1.4 |
| Piscivores | 0.1 | 0 | 0 | 0 |
| No. species in sample of 800 birds | 120 | 100 | 86 | 94 |
| Comparison with unlogged ($\chi^2$ value) | | 53.5 | 142.6 | 92.2 |
| *P* | | <0.001 | <0.001 | <0.001 |

Sample sizes are 800 birds except in the 12–13-year-old forest, where sample size is 500 individual birds. In this latter case the number of species in a sample of 800 birds is predicted from a logarithmic model of species accumulation over time. $\chi^2$ tests are performed on original data (frugivores combined, insectivore-frugivores combined, carnivores combined). Table excludes aerial insectivores (swifts and swallows) and migrant species.

*Sources*: Johns (1989b); A. Grieser Johns (unpublished data).

excluding bats and murids) a high percentage are nocturnal or active both at night and during the day (49% of 49 species at Makokou, Gabon, and 59% of 54 species at the Tekam Forest Reserve, peninsular Malaysia) (Emmons *et al.* 1983, Johns 1989a). Many large mammals are rare and many occupy large home ranges. All these factors make the forest mammal community extremely difficult to study except by indirect means.

Comparative surveys of large mammal communities in unlogged and logged forests are restricted to two sites in Malaysian dipterocarp forests (Table 6.5). Frugivores and folivores have been studied at Lopé, Gabon (Table 6.6).

In Malaysia, there are no definable changes in numbers of species between unlogged and logged[1-18:51] forest, but changes do occur in their abundances. Overall encounter rates fall following logging but in each case this is largely due to extreme decreases in single taxa (giant flying squirrels *Petaurista* at Tekam and mousedeer *Tragulus* at Ulu Segama). There is a tendency for terrestrial folivores and terrestrial frugivore-folivores to become more abundant in logged forest. The abundance of carnivores tends to peak 6–10 years after logging, prior to re-establishment of the lower canopy in regenerating forest.

At the Gabonese site there is a significant overall difference in species abundances in the more recently logged[3-5:11] forest, particularly among some primates, but no significant difference between unlogged and old logged[10-15:11] forest. Low mammal densities at the Gabonese site meant that differences in individual species densities between sites could generally not be tested statistically. Only chimpanzees *Pan troglodytes* showed a significant drop in density following logging. White (1992) suggests, however, that inclusion of data from other sites around Lopé would demonstrate significant drops in density among several more primates and increases in the density of gorillas and elephants following logging. This is largely a reflection on logging intensity: a lack of clear trends at Lopé perhaps reflected the very low timber extraction levels.

### Attributes of species assemblages
#### Describing changes within assemblages
There are no easily defined techniques for demonstrating the extent to which tropical forest communities change under various types of forest management practice. To a large extent this reflects the complexity of a dynamic ecosystem. This is emphasized by Putman (1994) who interprets data from the Tekam Forest Reserve, peninsular Malaysia, to suggest that monitoring different 'stability functions' (*sensu* Orians 1975) would give rather different results. In this example, the vertebrate community shows little inertia (the structure of the community was substantially changed by the felling event),

low constancy (retention of the structure of the community in terms of feeding guilds, etc., but substantial changes in the relative abundance of species), but reasonable resilience (an ability to recover to the original state).

Ideally, management planning for biodiversity retention would incorporate information on the complete forest community and analysis would be possible within established frameworks of community and diversity theory (summarized in Putman 1994). In practice, it will rely on easily obtainable statistics concerning a few targeted species or species assemblages. Analysis is commonly restricted to large mammals and birds, but occasionally it includes other large species arrays such as trees, butterflies, hawkmoths or other groups which contain large species arrays and for which taxonomy is not problematic (e.g. Howard 1991). Although not optimal, information of this type can be important in determining changes occurring following logging and in predicting the likelihood of the forest community recovering to its former state.

### Species richness and diversity indices

As is often overlooked, species richness and diversity are not necessarily the same thing. Species richness refers to the actual numbers of species in a given habitat, diversity is a function of the numbers of species and their relative abundances (the species-abundance distribution). Increasing diversity as recorded by a diversity index does not correlate with increasing species richness. For example, the bird community in the Tekam Forest Reserve, peninsular Malaysia, exhibited an increase in diversity of up to 10% in logged$^{(\geq 6:51)}$ forests (as expressed by the Shannon–Wiener Index) but a 20–37% decrease in species richness (as determined from population samples).

Species richness can be estimated from cumulative abundance distributions, but this becomes more accurate the larger the sample size (both in terms of numbers of individuals and the total area sampled). Only if very large comparative samples from unlogged and logged forests continue to be significantly different can it reasonably be assumed that differences are real. Small samples are affected by changes in the species-abundance distribution: rare species may become rarer and fewer are recorded in the sample.

Differences between habitats are commonly examined using simple diversity indices (e.g. Beehler *et al.* 1987). Diversity among animals commonly correlates with diversity of habitat structure. For example, bird species diversity has commonly, although not always, been correlated with vegetational height diversity or other aspects of foliage structure and volume (Willson 1974; c.f. Lovejoy 1974). Logging may act to increase habitat heterogeneity on a local scale. Small study plots in logged forest often contain a higher diversity of successional vegetation stages than is typical of similarly sized plots in

Table 6.5. *Mammals species recorded in unlogged and logged Malaysian dipterocarp forests*

| | No. species recorded (groups encountered/100 km) | | | | | | |
|---|---|---|---|---|---|---|---|
| | Tekam Forest Reserve, peninsular Malaysia | | | | Ulu Segama Forest Reserve, Sabah | | |
| Feeding guild and family | Unlogged | 1- to 6-year-old logged | 7- to 12-year-old logged | 13- to 18-year-old logged | Unlogged | 6-year-old logged | 12-year-old logged |
| **Folivores** | | | | | | | |
| Cynocephalidae[a] | 1 (1) | 0 | 1 (1) | 0 | 1 (4) | 0 | 0 |
| Elephantidae[b] | 1 (3) | 1 (4) | 1 (4) | 1 (2) | 1 (1) | 1 (3) | 1 (12) |
| Tapiridae[a] | 1 (1) | 1 (6) | 1 (6) | 1 (7) | — | — | 0 |
| Bovidae | 0 | 1 (1) | 1 (2) | 0 | 0 | 1 (1) | |
| **Frugivores** | | | | | | | |
| Pongidae | — | — | — | — | 1 (5) | 1 (6) | 1 (3) |
| Sciurinae | 8 (111) | 9 (80) | 8 (110) | 7 (106) | 10 (23) | 5 (16) | 4 (16) |
| **Frugivore-folivores** | | | | | | | |
| Cercopithecidae | 3 (37) | 4 (57) | 4 (57) | 4 (68) | 5 (37) | 3 (28) | 4 (38) |
| Hylobatidae | 1 (26) | 1 (18) | 1 (30) | 1 (40) | 1 (15) | 1 (23) | 1 (13) |
| Petauristinae[a] | 4 (393) | 3 (144) | 3 (50) | 3 (27) | 2 (27) | 2 (15) | 1 (10) |
| Hystricidae[a] | 1 (1) | 0 | 1 (3) | 1 (3) | 3 (2) | 1 (1) | 1 (2) |
| Suidae[b] | 2 (10) | 1 (5) | 1 (12) | 1 (8) | 1 (14) | 1 (16) | 1 (21) |
| Tragulidae[a] | 1 (1) | 1 (3) | 2 (10) | 2 (10) | 2 (139) | 0 | 1 (12) |
| Cervidae[b] | 2 (1) | 2 (6) | 2 (9) | 2 (1) | 2 (61) | 2 (12) | 2 (21) |
| **Frugivore-faunivores** | | | | | | | |
| Lorisidae[a] | 1 (80) | 1 (40) | 1 (43) | 1 (37) | 1 (2) | 0 | 0 |
| Ursidae | 1 (1) | 0 | 1 (3) | 1 (1) | 1 (1) | 0 | 1 (5) |
| Viverridae[a] | 0 | 3 (9) | 5 (33) | 3 (13) | 6 (41) | 3 (6) | 1 (10) |

| | | | | | | | |
|---|---|---|---|---|---|---|---|
| **Insectivores** | | | | | | | |
| Erinaceidae[a] | 1 (3) | 0 | 1 (1) | 0 | 1 (2) | 0 | 0 |
| Tupaiidae[b] | 3 (15) | 1 (9) | 2 (5) | 1 (13) | 3 (5) | 2 (7) | 2 (7) |
| Tarsiidae[a] | — | — | — | — | 1 (8) | 0 | 0 |
| Manidae | 0 | 1 (1) | 0 | 0 | 0 | 0 | 0 |
| **Carnivores** | | | | | | | |
| Canidae | (1) | 1 (1) | 0 | 1 (2) | — | — | — |
| Mustelidae | 2 (1) | 0 | 0 | 1 (3) | 4 (3) | 2 (7) | 1 (5) |
| Felidae[a] | 3 (3) | 6 (13) | 4 (11) | 3 (3) | 2 (1) | 2 (3) | 0 |
| Total species | 37 | 37 | 40 | 34 | 48 | 27 | 22 |
| Total encounters/100 km | 689 | 399 | 400 | 353 | 391 | 144 | 175 |
| **Survey distance (km)** | | | | | | | |
| Diurnal | 200 | 105 | 90 | 90 | 185 | 67 | 43 |
| Nocturnal | 50 | 30 | 30 | 30 | 44 | 20 | 10 |
| **Comparison with unlogged (encounter rates):** | | | | | | | |
| $\chi^2$ value[c] | | 15.0 | 61.9 | 68.5 | | 30.4 | 19.4 |
| $P$ | | <0.05 | <0.001 | <0.001 | <0.001 | <0.001 | <0.01 |

[a] Nocturnal: group encounter rate is calculated from nocturnal surveys.
[b] Active during day and night: encounter rate is calculated from both diurnal and nocturnal surveys.
[c] $\chi^2$ value combines all families within feeding guilds.
*Source*: A. Grieser Johns (unpublished data).

Table 6.6. *Abundance of selected mammals in unlogged and logged forest in Gabon*

| Family and species | Unlogged | Individuals/km² | |
| | | 3- to 5-year-old logged | 10- to 15-year-old logged |
|---|---|---|---|
| **Folivores** | | | |
| Elephantidae | | | |
| *Loxodonta africana* | 2.7 | 1.0 | 1.1 |
| Bovidae | | | |
| *Syncerus caffer* | 0.1 | 0 | 1.1 |
| **Frugivores** | | | |
| Scuiridae | | | |
| 'Squirrels' | 8.4 | 10.0 | 12.4 |
| **Frugivore-folivores** | | | |
| Cercopithecidae | | | |
| *Cercopithecus nictitans* | 23.9 | 7.0 | 25.7 |
| *C. pongonius* | 5.4 | 2.3 | 5.8 |
| *C. cephus* | 0.1 | 0 | 0 |
| *Cercocebus albigena* | 10.0 | 2.8 | 9.2 |
| *Colobus satanas* | 12.7 | 15.3 | 13.2 |
| *Mandrillus sphinx* | 1.8 | 3.8 | 5.3 |
| Pongidae | | | |
| *Gorilla gorilla* | 1.4 | 0.9 | 0.8 |
| *Pan troglodytes* | 0.5 | 0 | 0.3 |
| Suidae | | | |
| *Potamochoerus porcus* | 1.5 | 5.4 | 3.3 |
| Cephalophidae[a] | | | |
| *Cephalophos monticola* | 0.9 | 1.6 | 0.9 |
| *C. sylvicultor* | 4.5 | 0 | 0.3 |
| 'Red duikers' | 15.7 | 4.9 | 3.5 |
| Comparison with unlogged ($\chi^2$ value)[b] | | 16.0 | 2.57 |
| *P* | | <0.001 | >0.05 |

Logging at this site concentrated almost exclusively on *Aucoumea klaineana*, which made up 64% of trees cut; damage levels were 11%.
[a]Estimates derived from dung sampling.
[b]Comparison of original data on encounter rates over ten censuses (all species within feeding guilds combined).
*Source*: Adapted from White (1992).

unlogged forest and a higher diversity of bird species is often recorded within them (Table 6.7). However, the use of diversity indices overlooks the fact that different subsets of species are reacting to disturbance in different ways (Karr & Roth 1971). Although diversity is higher this does not preclude the decline or absence of mature forest specialists. Diversity indices are of limited value in assessing the extent to which species assemblages change following logging and implications for improved forest management.

Equitability tends to increase consistent with a greater evenness of species abundances in logged forest (fewer rare species are recorded). However, low equitability may be recorded where single species become over-dominant in disturbed forest. For example, low equitability recorded in disturbed forests of the Eastern Ghats is explained by the fact that three bird species accounted for more than half the individuals censused in these forests (Beehler *et al.* 1987).

Table 6.7. *Diversity indices for bird assemblages in undisturbed and disturbed forest*

| Area and habitat type | Shannon–Wiener diversity index H′ | Equitability E |
|---|---|---|
| Eastern Ghats, India[a] | | |
| Lightly disturbed | 2.91 | 0.57 |
| Moderately disturbed | 3.06–3.10 | 0.50–0.55 |
| Heavily disturbed | 2.69 | 0.40 |
| Bia Forest, Ghana[b] | | |
| Unlogged | 4.04 | 0.86 |
| 2-year-old logged | 4.06 | 0.86 |
| 10-year-old logged | 4.08 | 0.87 |
| Tekam Forest Reserve, peninsular Malaysia[c] | | |
| Unlogged | 5.25 | 0.83 |
| 1 year-old logged | 3.94 | 0.68 |
| 7 year-old logged | 5.45 | 0.85 |
| 13 year-old logged | 5.31 | 0.84 |

[a]Lightly and moderately disturbed forest influenced by cutting of firewood and burning of the forest understorey, heavily disturbed forest consists of natural overstorey with underplanted coffee.
[b]Damage level during logging of 15%.
[c]Damage level during logging of 51%.
*Sources*: [a]Beehler *et al.* (1987); [b]Holbech (1992); [c]A. Grieser Johns (unpublished data).

Changes in equitability may therefore occur in either direction consistent with changes in the relative abundance of species.

### Indices of similarity

The most consistent effect of environmental disturbance is to alter the relative abundance of species, although there is no consistency in the form of alteration. Comparative species abundances may offer perhaps the most effective assessment of how different mature and logged forests actually are. Similarity is most often measured using Indices of Overlap (e.g. Horn 1966), which compare the relative numbers of individuals of named species.

Patchiness of species distributions means that there is variation in species abundances even between closely adjacent areas of unlogged forest. Horn's Index of Overlap gives a value of 0.72 comparing bird populations in two unlogged forest sites in Sabah. Also, due to transience or extensive ranging patterns of many larger animals, there is variation in species abundances over time within an area of unlogged forest. At the Tekam Forest Reserve, peninsular Malaysia, Horn's Index of Overlap gives a value of 0.68 comparing bird populations between months (Table 6.8).

In general, a high level of similarity would be expected in the above comparisons. Similarity would be expected to become less as a reflection of the degree of change to the forest habitat following logging, but to increase again parallel to the success of forest regeneration. Unfortunately, data are not yet available to determine if similarity indices regain their original value in fully regenerated forests prior to their being re-logged. A comparison of similarity of bird species abundances (as measured by Horn's Index of Overlap) demonstrates that even older logged forests can be highly dissimilar to unlogged forest (Table 6.8). This reflects the changes that occur in the relative abundance of species and which persist into old regenerating forests.

Statistics of this type are highly sensitive to habitat parameters and different effects may be difficult to isolate. An important influencing factor is edge effect, particularly if unlogged forest sites studied are close to forest/agriculture interfaces. In the West Mengo Forest Reserve, Uganda, understorey birds sampled by mist-netting gave overlap values of 0.87–0.93 where a 450 hectare unlogged forest was compared with four logged[25–30: > 50] forest sites (Dranzoa & Johns 1992). This high degree of overlap is probably due to degeneration of the small unlogged forest reserve through edge effects rather than recovery of the avifauna of logged forest (which contained an average of only 29% of the number of trees and 63% of the basal area of the unlogged forest).

In another example, comparisons of the avifauna within a range of unlogged and logged[9–48: > 50] compartments of the Budongo Forest Reserve, Uganda, gave overlap values of 0.72–0.92 and were not correlated in any way with logging history (I. Owiunji & A. Plumptre, pers. comm.). In addition to edge effects, which may again be influencing results from the unlogged sites, there was variation in habitat topography, vegetation type and bird species composition between the unlogged sites, which were 30 km apart. All these factors may mask changes in the relative abundances of species due to logging.

Even if distant from forest/agricultural interfaces, samples from unlogged forest may be affected by the presence of natural disturbed vegetation types such as floodplain or riverine forests. In an unlogged study site at M'Passa, Gabon, which contained large areas of riverine forest, one-third of the total bird density recorded was made up of insectivorous-frugivorous greenbuls, with one species *Andropadus latirostris* occurring at a density of 500 individuals/km$^2$ (Brosset 1990).

In any of the above examples, monitoring change through individual species abundances could be misleading. It is important that the unlogged forest site with which logged forests are to be compared is completely uninfluenced by edge effects from either natural or human-induced sources.

Table 6.8. *Similarity indices for bird assemblages (comparison with unlogged forest)*

| Area and habitat type | Tree loss during logging (%) | Horn's index of overlap |
|---|---|---|
| Tekam Forest Reserve, peninsular Malaysia[c] | | |
| Unlogged | | 0.68[a] |
| 1-year-old logged | 51 | 0.51 |
| 7-year-old logged | 51 | 0.07 |
| 13-year-old logged | 51 | 0 |
| Ulu Segama Forest Reserve, Sabah[c] | | |
| Unlogged | | 0.72[b] |
| 6-year-old logged | 57 | 0.34 |
| 12-year-old logged | 58 | 0.12 |
| Kibale Forest, Uganda[d] | | |
| Unlogged | | — |
| 23-year-old logged | ? > 50 | 0.14–0.19 |

[a]Comparison of five data sets collected in five dry-season months at the same site.
[b]Comparison of data from two unlogged forest sites 15 km apart.
*Sources*: [c]A. Grieser Johns (unpublished data); [d]Dranzoa (1995).

*Indicator species*

Definition of an 'indicator species' (a species whose abundance can be used to predict some characteristic of the ecosystem) is usually difficult. Defining species to act as indicators of logging effects is especially difficult since these effects can be quite subtle in comparison with environmental disturbances such as commercial hunting or forest clearance.

Attempts to define attributes of species which permit or promote resilience in the face of habitat disturbance have been limited to isolated ecosystems (Terborgh & Winter 1980). On forest islands created after the flooding of large land areas, species rarity is an important factor. Populations become vulnerable to unpredictable environmental fluctuations when their populations are at very low levels (Willis 1974). However, rare species are not consistently adversely affected by logging. Combinations of body size and broad feeding niche can result in relatively consistent adverse reactions to logging in some species groups (e.g. large-bodied frugivorous primates) but there are often exceptions (Johns & Skorupa 1987; Fimbel 1994).

Precise knowledge of a species's niche can help predict response to logging, but niches of tropical forest animals are often overlapping rather than discrete. A species may be capable of expanding its foraging niche to fill any 'gaps' without the acquisition of new behaviour. In isolated ecosystems, species occupying narrow niches have been shown to have a statistically higher probability of elimination (e.g. Karr 1982). However, no correlation exists between niche breadth and sensitivity to logging because species exhibiting opposite reactions have equally narrow niches. For example, at Ponta da Castanha, Brazil, narrow niche breadths (defined by Levins's (1968) Index of Niche Breadth) were shared by mature forest specialists such as scythbills *Campylorhamphus* which persist poorly in disturbed forest and by finches Fringillidae which invade heavily disturbed forest (Johns 1991b).

*Indicators of logging impacts*

For practical reasons, the choice of indicator species has been heavily biased towards mammals and birds, which are unlikely to reflect changes at a range of temporal, spatial and organizational levels (Landres *et al.* 1988). Furthermore, many larger vertebrates are able to respond quite rapidly to spatial and temporal factors.

Hornbills are large-bodied frugivores with often quite specialized diets and have been considered potential indicators of the success of forest regeneration and retention of tree diversity. However, they are very wide ranging animals and their abundance fluctuates more in response to seasonal changes in food supply than to local habitat features *per se*. At the Tekam Forest Reserve, peninsular Malaysia, monthly fluctuations in the abundance of hornbills were not significantly different before and after logging[0-1:51] or between unlogged forest samples and samples from a series of logged[1-18:51] forest sites (Fig. 6.1). This remained the case where the helmeted hornbill *Rhinoplax vigil* was examined separately; this species is considered to be a fig specialist and the most likely potential indicator species.

In some cases, demonstrable absences of individual vertebrate species with specialized niches may be indicative of local environmental changes. For example, an absence of piscivorous kingfishers reflects silting or other logging damage to watercourses, or an absence of large terrestrial birds reflects the levels of soil compaction and loss of leaf litter invertebrates (in the absence of hunting by humans). Causal relationships may be proven, but they are not indicative of wider ecosystem processes. Comparisons between different areas often show rather different responses by the same species under similar levels of environmental change (see Chapter 7), suggesting that species may be responding to very precise features of habitat or food supply.

A better approach may be to look within the invertebrates for indicator groups. For example, Holloway (1984b) suggests that abundances of moths of

Fig. 6.1. Monthly variation in encounter rates of all hornbill species ($\bar{x}$ + SE) and the helmeted hornbill *Rhinoplax vigil* ($\bar{x}$), Tekam Forest Reserve, peninsular Malaysia. Lines link data from area C13C collected before and after the onset of felling. Other data points are from subsequent surveys of C13C and surveys at other sites in the timber concession. (Source: A. Grieser Johns unpublished data.)

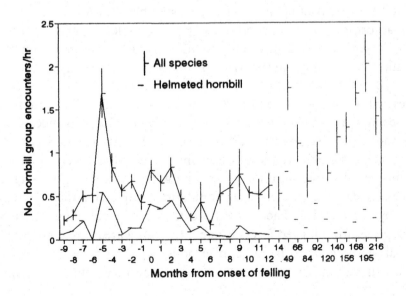

the family Arctiidae are specific to levels of lowland forest disturbance and regeneration processes, and abundances of the Geometridae will be similarly useful in montane forests. These ideas have yet to be tested.

*Flagship species as indicator species*
'Flagship' species are typically large and conspicuous vertebrates well known to the public in industrialized countries. They share no common ecological trait, but it is assumed that what benefits them, in terms of habitat conservation measures, benefits the ecosystem as a whole. There is a tendency to use flagship species as indicator species (e.g. healthy populations of tigers should indicate successful retention of tiger habitat and associated species).

In some cases this may be entirely justified. The northern spotted owl *Strix occidentalis caurina* of the Pacific North-West United States is an example. This species may indeed be an appropriate indicator of the retention of habitat quality in the coniferous forests (Hunter 1990). However, densities of large owls, including the very large powerful owl *Ninox strenua*, were not reduced in a mosaic of logged[0.5-23:100] subtropical forests in southern Australia (Kavanagh & Bamkin 1995). The powerful owl may fulfill a role as a flagship species but is not a good indicator species. The choice of indicator species needs to be specific to particular forest types and possible particular locations within the forest types.

### Summary
Geographical differences exist in the structure of rain forest communities such that there are no overall linear relationships between logging intensity and the extent of change in species assemblages. Species responses also vary according to local conditions. For example, changes in litter invertebrate assemblages are directed primarily towards local microclimatic changes rather than changes in the structure or composition of the forest as a whole.

Changes within species assemblages can be described by a variety of statistical techniques, which often provide conflicting results. This reflects the differing responses to disturbance of different ecosystem parameters. Studies of vertebrate species assemblages show little change in overall numbers of species or the frequency of species in different feeding guilds as a result of logging. Species richness may tend to decrease in population samples, but this is heavily influenced by sample size. There is little evidence of local elimination of vertebrate species from logged forest where large study areas are considered.

Species diversity tends to increase in response to changes in the species-abundance distribution. The relative abundance of species within assemblages can change very markedly following logging. Indices of overlap usually indicate high levels of dissimilarity between unlogged and logged forest samples. This dissimilarity persists into regenerating logged forest. Studies have not yet continued for long enough to determine whether recovery to unlogged forest values can eventually occur.

There has been little success in determining indicator species for use in forest management planning since logging impacts can be rather subtle compared with natural spatial variation in species distributions. Vertebrates are generally too mobile or too adaptable to changing environmental conditions to make good indicators of site-specific logging impacts. Suitable indicator taxa are more likely to be found among invertebrate groups, but these have been little investigated.

# 7

## Using ecological data in forest management planning

### Introduction

The application of ecological data on the effects of timber logging on biodiversity to forest planning and *in situ* forest management is at a very early stage of development. There are two principal reasons for this.

First, the level of training of forestry staff in many tropical countries is often very low (Palmer & Synnott 1992). Field staff often lack basic technical manuals: research is outdated and poorly disseminated. Staff are not familiar with the range of forest management techniques at their disposal and have very little knowledge or appreciation of the relevance of biodiversity conservation. Even if basic principles of biodiversity conservation are included within national forest planning strategies, to be applied effectively these require both research programmes to determine specifically the type and extent of any interventions required and training of forestry staff to implement them.

Second, research studies of the effects of timber logging on biodiversity have generally not collected data in a form that can be applied to *in situ* forest management. Most research has been academically oriented with the implications for forest management considered only after the study has been completed. The design of most field studies has been particularly limiting: most are short-term comparisons of sites of different logging histories. To make an effective contribution to forest management planning, studies need to be long-term and directed towards the identification of cost-effective techniques for enhancing the biological value of regenerating forest.

This chapter considers the reliability of information produced by different types of ecological study and their potential contribution to forest management practice. The next chapter considers the forms of intervention that might be applied.

### The problem of spatial heterogeneity

A principal problem facing tropical forest management is the high degree of spatial heterogeneity. Within primary forest, variation in the relative density of subhabitats and their associated plant and animal species can be considerable over even quite short distances. This depends partly on local historical factors, including the pattern of different regeneration phases in gaps of different ages, but perhaps primarily on local edaphic conditions. The result is considerable local variation in the densities of plant and animal species (e.g. Braithwaite *et al.* 1984).

The influence of natural patchiness carries over into species distribution within production forests. Studies able to assess this phenomenon objectively have generally found that patchiness within logging concessions relates more to historical factors, elevation and soil fertility than to logging intensity or damage levels (e.g. Johns 1989a; Kavanagh & Bamkin 1994). In effect, variation due to spatial heterogeneity can be greater than variation due to the effects of timber extraction, at least up to damage levels of 50% or thereabouts.

There are two aspects of spatial heterogeneity to be considered: variations in forest structure and variation in species composition.

#### Structural variation

The effects of logging on the structure of forest vegetation needs to be considered on various scales. In the dry forests of western Madagascar, changes in vegetational structure caused by logging[8–9: <50] were significant within small study plots (Ganzhorn *et al.* 1990). However, where larger areas were considered, variation caused by logging was mostly smaller than the variation of forest structure due to natural causes. In the context of forest management units, the impact of logging on the forest vegetation was considered secondary to natural variation and forest patchiness.

Whether or not spatial heterogeneity in structural variables may be correlated with logging impacts is largely a question of damage levels incurred, however. The Madagascar example involves very low (unfortunately not quantified) damage levels incurred during removal of $<10 \, \text{m}^3/\text{ha}$ of timber. Where greater damage levels are incurred, variation in forest structure on a large scale may be correlated with logging impacts (e.g. in logged[1–8: >50] forest of West Kalimantan: Cannon *et al.* 1994).

#### Variation in species composition

Spatial heterogeneity in species composition may be expressed either in patchy distribution of species or in local variation in their relative abundances. In the former case, ranges may be discontinuous through a forest habitat for reasons not easily explained on the basis of broad habitat features. For example, lacunae in the local distribution of six species of large squirrels results in different species compositions at each of four points along a 16 km transect through logged[1-18:51] dipterocarp forest (Table 7.1).

Differences in relative abundance can be equally unpredictable. In the Budongo forest of Uganda, variation in bird populations between unlogged and logged[9-48:>50] forest sites could not be correlated with logging history or other linked variables (I. Owiunji & A. Plumptre, pers. comm.). Both bird and primate populations varied spatially to such an extent that studies of the impact of logging on the species populations would give quite different results depending on which unlogged and logged compartments were actually compared (Plumptre *et al.* 1994). Similarly, primate densities in the Tekam Forest Reserve, peninsular Malaysia, have been shown to vary more between different sites (along a 20-km transect) than at the same site over a logging[0-18:51] process (Johns 1989a; Grieser Johns & Grieser Johns 1995).

Spatial heterogeneity is typically not controlled for in studies of the effects of forest management practices on biodiversity. Most studies compare unlogged forest with adjacent or nearby areas of forest of varying management histories (i.e. inter-site comparisons). In this type of study it has to be assumed that the

Table 7.1. *Presence/absence of giant squirrel and giant flying-squirrel species along a 16 km transect in the Tekam Forest Reserve, peninsular Malaysia*

|  | Points (km) along transect | | | |
| --- | --- | --- | --- | --- |
| Species | 0[a] | 6.5[b] | 11[c] | 16[d] |
| Sciurinae (diurnal squirrels) | | | | |
| *Ratufa affinis* |  | x | x | x |
| *R. bicolor* | x | x | x | x |
| *Callosciurus prevostii* |  | x | x | x |
| Petauristinae (flying squirrels) | | | | |
| *Petaurista petaurista* | x | x | x | x |
| *P. elegans* | x |  |  | x |
| *Aeromys tephromelas* | x |  | x | x |

Damage level during logging was 51%.
x, present.
[a]Forest sampled 1, 7 and 13 years post-logging.
[b]Forest sampled 2, 8 and 14 years post-logging.
[c]Forest sampled 4, 10 and 16 years post-logging.
[d]Forest sampled 6, 12 and 18 years post-logging.
*Source*: A. Grieser Johns (unpublished data).

biological communities of the different sites were identical before felling events, and thus that any differences observed are due to environmental or other changes caused by the felling. In other words, it is assumed that the effects of logging are greater than natural spatial variation. From the examples given above, this underpinning assumption must be called into question.

Clearly, the optimal study design is to monitor changes in plant and animal populations at the same site over a complete logging cycle. This has the disadvantage of taking a very long time. Only a few studies have begun prior to logging of a study area and continued for a period thereafter (Tekam Forest Reserve, peninsular Malaysia[0-13:51]: Johns 1989a, Grieser Johns & Grieser Johns 1995; Queensland[0-2:16-23]: Crome & Moore 1989; Lopé, Gabon[0-1:11]: White 1992).

A typical justification for inter-site comparative studies is that results are required over the short term to indicate whether changes in management practices are required and that these should be applied quickly as part of overall forest planning (e.g. ODA 1995). This begs two questions. First, at what level of information can inter-site comparisons be considered reliable and can such data be used to define requirements for specific interventions? Second, how applicable are results from single sites or a few sites to a wider forest estate?

### Inter-site comparisons: information content

Currently, there is only a single dataset large enough to enable an objective comparison between the 'real' effects of logging and the 'predicted' effects of logging. The first refers to data obtained from monitoring a single site before, during and after a felling operation (i.e. the effects of spatial heterogeneity are controlled for). The second refers to comparative data obtained from inter-site comparisons where the original communities are assumed to be the same (i.e. the effects of spatial heterogeneity are not controlled for).

This study is located in the Tekam Forest Reserve of peninsular Malaysia and has monitored a cohort of sites logged between 1975 and 1981. In one of these sites (C13C) data were collected for 18 months prior to the onset of logging. Results from C13C which has so far been monitored before, during and up to 13 years post-logging document the 'real' effect of logging. Comparisons of data collected from this site prior to logging with data collected from three other sites (C5A, C1A and C2) logged between 1 and 18 years previously gives a 'predicted' effect of logging akin to that obtained from the more typical inter-site comparative studies. The extent to which there is agreement between 'real' and 'predicted' effects can be examined on three levels: the responses of species assemblages (birds, mammals, etc.), the responses of guilds of species of similar feeding habits, and the responses of individual species.

*Species assemblages*

Effects of logging on species assemblages are typically expressed in terms of changes in species richness or diversity, and occasionally in terms of the level of similarity in species abundances between unlogged and logged forests. The 'real' and 'predicted' changes in these parameters can be illustrated from data collected on the bird community at Tekam (Fig. 7.1). There is general

Fig. 7.1. Changes in a bird species assemblage as recorded by species richness (a), species diversity (b), and Horn's index of overlap (c) at Tekam Forest Reserve, peninsular Malaysia. Lines link data from site C13C. Other data points are from sites C5A (cross), C1A (star) and C2 (open square). (Source: A. Grieser Johns unpublished data.)

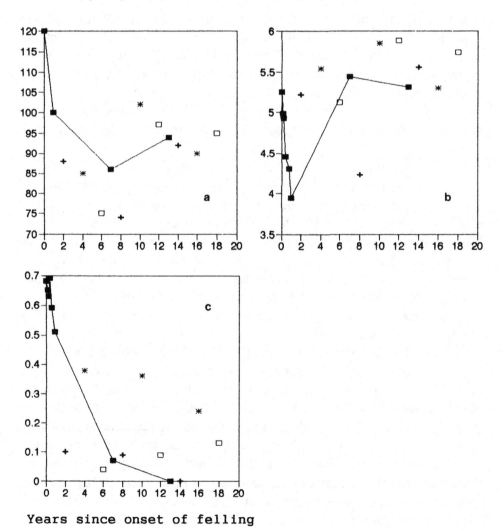

**Years since onset of felling**

agreement, suggesting that inter-site comparisons provide an approximation of the 'real' changes and thus are useful at this level of analysis.

However, as mentioned earlier, analyses of this type are heavily influenced by sampling techniques and sample sizes. Although small samples record a drop in species richness, for example, this reflects the different shape of species-accumulation curves. Since the curves converge at higher sample sizes, the greater the sampling effort the less likely that differences will exist between unlogged and logged forest. The form of comparison employed, be it 'real' or 'predicted' is in this case less important than the sampling effort.

Where large samples are applied either technique should determine the extent to which changes are actually occurring within a species assemblage. If drops in species-richness are indicated, for example, then a need for further study is indicated.

### Guilds

Many logging studies have looked at differences in densities of species or individuals within feeding guilds. For example, studies typically show drops in the relative frequency of understorey and terrestrial insectivorous mammals and birds in logged forest compared with increases in the frequency of observation of generalist species associated with edge habitat.

Generalities of this type can be demonstrated both from inter-site and intra-site monitoring (Table 7.2). However, the significance of differences is greater where inter-site comparisons are involved, which may be a reflection of undefined variation in the local effects of logging.

Inter-site heterogeneity may make it difficult to determine reasons for changing inter-guild relationships over time. For example, some inter-site comparative studies demonstrate that bark-gleaning insectivorous birds are less abundant in logged[8;57] forests, perhaps due to a loss of snags providing foraging substrata and breeding sites (e.g. Lambert 1992). Others do not (e.g. Grieser Johns 1996). Monitoring at Tekam demonstrates that there may be periods in the regeneration cycle when there are few snags but that continuing mortality of residual large trees ensures that this is an irregular or temporary phenomenon on a small spatial scale. Such changes are rather subtle and may not be detected simply by comparing different logged forest sites of different logging histories.

From overall changes within guilds it may be possible to hazard a guess as to the reaction of member species to logging at certain intensities. If terrestrial insectivores as a group are 50% less abundant in logged forest of a particular regeneration stage then it might be assumed that member species are adversely affected to approximately that degree. This is notoriously unreliable, however, as members of the same guild can react in completely different ways to logging. Understorey flycatchers are an example: these have been reported to be particularly prone to logging (for reasons such as changes in food supply, alteration of light conditions in the understorey and introduced competition with aerial insectivores such as treeswifts). However, a few understorey flycatcher species are actually edge habitat specialists and tend to increase in abundance in logged forest. Changes in the frequency of species within guilds is thus not very useful information for management purposes.

Table 7.2. *A comparison of feeding guild membership (% total individuals) within a bird species assemblage, Tekam Forest Reserve, peninsular Malaysia*

| | Years post-logging | | | | |
| | Area C13C | | | Area C5A | Area C2 |
| Feeding guild | 0 | 7 | 13 | 8 | 12 |
|---|---|---|---|---|---|
| Frugivores | | | | | |
| Terrestrial | 0 | 0 | 0 | 0 | 0.3 |
| Arboreal | 10.6 | 14.9 | 11.9 | 31.7 | 22.0 |
| Faunivore-frugivores | 11.5 | 5.8 | 6.6 | 4.3 | 4.5 |
| Insectivore-frugivores | | | | | |
| Terrestrial | 0.6 | 0 | 0.2 | 0 | 0 |
| Arboreal | 15.1 | 39.2 | 31.0 | 30.6 | 23.9 |
| Insectivore-nectarivores | 3.9 | 4.5 | 8.9 | 3.9 | 2.5 |
| Insectivores | | | | | |
| Terrestrial | 3.6 | 0 | 2.4 | 0.1 | 0.5 |
| Bark-gleaning | 5.3 | 3.0 | 3.8 | 1.1 | 3.0 |
| Foliage-gleaning | 31.1 | 28.5 | 28.4 | 14.6 | 33.1 |
| Sallying | 16.2 | 2.0 | 5.2 | 10.8 | 6.5 |
| Carnivores | | | | | |
| Raptors | 2.0 | 2.0 | 1.4 | 2.9 | 3.5 |
| Piscivores | 0.1 | 0 | 0 | 0 | 0.3 |
| Comparison with C13C | | | | | |
| unlogged ($\chi^2$ value) | — | 43.2 | 84.6 | 242.7 | 131.6 |
| *P* | | <0.001 | <0.001 | <0.001 | <0.001 |

Sample size is 800 individuals in all cases except C13C 13 years post-logging where sample size is 500 individuals.
*Source*: A. Grieser Johns (unpublished data).

*Individual species abundances*

At the level of their effects on individual species the effects of logging operations are very site-specific. For example, since so much of the felling damage is random, some sites may lose a high percentage of certain non-timber fruit trees or other resources whereas others may not. This type of local variation is controlled by the use of large study sites. Unfortunately, most field studies are located within rather small study sites.

Monitoring at site C13C has suggested that a total of 62 bird species are consistently present in reduced abundance in logged[7&13:51] forest samples. Twenty species are present in increased abundance (excluding colonizing species not present before logging). If data from C13C are compared with samples from logged[1–18:51] forest at sites C5A, C1A and C2, 39 species show consistent decreases in abundance and 21 show consistent increases. The different techniques agree on 33 (53%) of the decreasing species and 15 (75%) of the increasing species.

Inter-site comparisons (even with quite large sample sizes) along a 16-km transect appear to differ considerably, particularly for species adversely affected by logging. It is quite likely that the greater the number of logged forest sites compared, the less the number of species showing consistent reactions to logging. These differences may be due both to local differences in initial population densities and specific local effects of the logging operations. The result is that inter-site comparisons can be quite misleading concerning the reactions of individual species to logging. Only monitoring studies will give an accurate picture of responses of individual species. This implies, however, that requirements for interventions to conserve species responding poorly to logging can be accurately formulated only as a result of monitoring studies and that they may be highly site specific.

## Applying local data to a wider forest estate

The problems inherent in interpreting ecological data for management purposes are clearly illustrated using an example from the Ulu Segama Forest Reserve, Sabah. The responses of individual bird species to logging[6:57] was initially studied in the southern section of the RBJ/1981 compartment by Johns (1989b). This study was later duplicated in the northern section of the same logged[8:57] compartment by Lambert (1990). The two studies compared results from the same logged compartment (using study plots <1 km apart at their closest point) with the same plot of unlogged forest (Danum Valley Conservation Area) some 15 km distant. Both studies used a combination of transect surveys and mist-netting to sample bird populations. Grieser Johns (1996) subsequently identified 15 species showing a significant decrease in the logged forest and 46 showing a significant increase. Using less objective criteria, Lambert (1992) suggests 37 species may be decreasing and 32 increasing in numbers. The two studies agree on only six decreasing species and 19 increasing species. In this case, variation not easily attributable to the form of logging is seen over a very short spatial and temporal distance.

If more widely separated studies are considered, the degree of divergence in responses of individual species generally becomes greater. First, results from logged[7&13:51] forest at C13C in the Tekam Forest Reserve may be compared with those from logged[25:>50] forest in the Pasoh Forest Reserve, 150 km to the south (Wong 1985). Agreement is reached on 13 (25%) of the species showing reduced abundances and two (12%) of the species showing increased abundances (comparing understorey species only). Second, results from C13C at Tekam may be compared with those collected in logged[6&12:57] forest at the Ulu Segama Forest Reserve, Sabah, where the avifauna is only slightly different in composition (Grieser Johns 1996). In this case, agreement is reached on eight (13%) of the species showing reduced abundances and 13 (65%) of the species showing increased abundances.

The level of agreement in species increasing in abundance reflects the stage of forest regeneration: the Tekam and Ulu Segama samples were very similar in this respect whereas the Pasoh sample was from a much later regeneration stage. The level of agreement in species decreasing in abundance is most important for management purposes and this is uniformly low. In effect, results from specific sites cannot be applied with any level of accuracy over a wider forest estate. It will not be possible to make management recommendations for a large forest area based on results from specific sites. It would be inefficient to recommend interventions to benefit particular species if consistently vulnerable species cannot be defined.

### Making management recommendations

In some cases, it may be possible to determine co-variance between vegetation and animal species assemblages through detailed inter-site comparisons. This has been the approach in more detailed logging studies such as that at the Budongo Forest, Uganda (Plumptre *et al.* 1994). Typically, however, such studies can account for only a small part of the total variation as they cannot account for original spatial heterogeneity in species populations.

Broad changes in biodiversity at the level of species assemblages or perhaps guilds may be predictable given a knowledge of damage levels caused by felling. However, information at this level translates more into broad forest policy than the definition of interventions to conserve biodiversity. For example, data concerning the increasing numbers and long-term persistence of generalist taxa may reinforce other indicators of the need for reduced-impact logging, such as the regeneration success of timber trees.

At the level of individual species the uncertainty principle inherent in biological systems comes into play. Only monitoring studies can record responses at the species level and can provide information concerning conservation-oriented interventions.

Two types of species-level information may be required for management purposes, depending on forest policy. First, the aim might be to determine if certain species are declining and if so what to do about it. Second, the aim might be to use biodiversity monitoring to impart information on biological sustainability, which may be needed for long-term management or marketing purposes (e.g. to allow stratified marketing under approved certification systems). A third potential aim might be to use biological data as an index of change of other factors, such as loss of tree diversity or regeneration success, microclimatic conditions, or anything else pertaining to sustainability of timber production. Realistically, however, it is probably more efficient to measure these directly.

### Management to retain biodiversity

If management is to be actively directed towards retaining biodiversity, then interventions can be applied to address specific ecological problems. Intervention is expensive and it is cost-effective to identify particular local requirements rather than to mark and preserve particular resource trees as part of general field practice.

The following scenario may be considered as an example. The minimum retention levels of snags of various size classes, and of old but still-living trees that will provide future snags, are critical to the foraging or breeding success of certain bark-gleaning birds and hole-nesting mammals or marsupials. In most logged forests these minimum levels are not reached: the species potentially affected can be shown to persist (apparent adverse effects may in most cases be shown to be due to spatial heterogeneity in original populations). Intervention to preserve snags may only be needed in compartments logged at particularly high intensities or where fires or fuelwood cutting have removed a percentage of the original snags. The need for specific snag retention will thus have to be interpreted on a felling block by felling block basis.

### Management for sustainability

Monitoring of ecological changes in regenerating forest can be a useful tool in determining the overall biological sustainability of the logging system employed. This type of information cannot be obtained from inter-site comparisons and thus relies on long-term monitoring programmes being put into place as a part of forest management practice.

Monitoring of ecological processes is rather complex and relies on specific decisions being made as to the importance or relevance of various ecological parameters. Different parameters tend to produce conflicting data.

Some parameters may be resilient to disturbance: changes will occur only after a certain level of disturbance has been reached and normality will quickly be restored as the level of disturbance falls off (Fig. 7.2a). Other parameters may show fragility to disturbance: rapid changes occur at even very low disturbance levels (Fig. 7.2b). Species richness is an example of a resilient characteristic and the relative abundance of species (i.e. similarity) an example of a fragile characteristic.

Some parameters may show limited resilience (Fig. 7.2c). While disturbance levels remain below a critical point (x), few changes occur to the system. If they exceed this point rapid changes occur and the level of disturbance will need to be reduced considerably before the system recovers to a point of resilience to

Fig. 7.2. Alternative responses to disturbance: (a) resilience; (b) fragility; (c) limited resilience; (d) alteration.

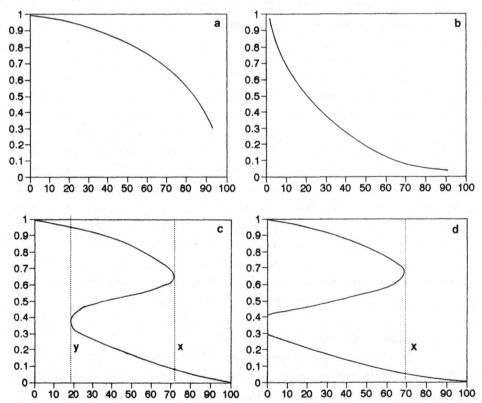

further disturbance (y). In the extreme case, exceeding the critical level of disturbance may cause a permanent change to the ecosystem (Fig. 7.2d). In effect, an arrested climax results.

Most ecosystem components will regenerate naturally to pre-logging levels given sufficient time between felling events. It is important, however, that ecological or environmental effects of logging that result in ecosystem parameters falling into the final category be identified. One example would be the level of canopy removal and associated herbaceous growth that encourages resident elephant populations, which would then inhibit tree regrowth through destructive browsing (some compartments of the Kibale Forest of Uganda demonstrate this effect). A second example would be the creation of environmental conditions that encourage the establishment of an invasive species, particularly an exotic (e.g. *Eupatorium* in Ghana: Hawthorne 1993).

Fig. 7.3. Changes in ecological parameters within a bird species assemblage following logging, Tekam Forest Reserve, peninsular Malaysia. Parameters are species diversity (H'), species richness (S) and Horn's index of overlap. All data refer to site C13C where basal area was reduced by 59% during logging. Increment after logging was approximately 1.5%/year. Points are connected in chronological sequence, beginning with unlogged forest (origin) sampled in January–June 1980, at three points during the logging process (July–December 1980), twice immediately after logging (March and June 1981), 7 years after the onset of felling (July 1987) and 13 years after the onset of felling (August 1993). (Source: A. Grieser Johns unpublished data.)

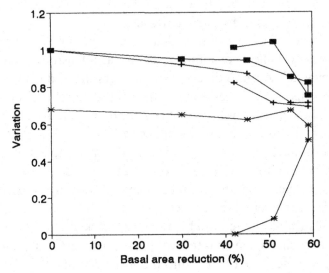

In broad terms, logging operations that result in ecosystem parameters falling into category d. are not biologically sustainable. Extensive restoration may be required to regain the original forest community. Where parameters fall into category c., sustainability can be assured only if re-logging does not occur until regeneration has passed point y. The time of re-logging is also critical where parameters fall into category b.

An example of this approach (Fig. 7.3) records resilience in bird species diversity and richness as opposed to fragility of bird species relative abundances. The importance of these parameters in determining biological sustainability is open to interpretation. A drop in species richness is probably an important criterion. A change in the relative abundance of species may be less important as it reflects changes in the vegetation as, for example towards a higher density of timber species, that may be important for economic reasons. However, re-logging before relative abundances reach pre-logging levels may eventually cause sufficient fluctuations in abundance that some species are lost through random environmental fluctuations. Monitoring would assess the changes over time and help determine when particular parameters become important in determining the response of the ecosystem as a whole.

### Summary

There is considerable spatial heterogeneity in plant and animal distribution within tropical forests. Ecological analyses of the effects of logging that are based on inter-site comparisons may indicate differences, particularly in species richness and composition, that might be due more to spatial heterogeneity than to the effects of forestry activities.

While it may be appropriate to use (detailed) inter-site comparisons to obtain an impression of gross effects of felling at particular intensities, such as effects on overall species richness and diversity, or changes in guild composition, inter-site comparisons are not reliable in documenting changes at the level of individual species. Thus they cannot provide reliable data on requirements for specific interventions to conserve individual species.

In general, reduced-impact logging procedures directed towards the improvement of timber regeneration will go a long way towards conserving biodiversity as a by-product, and this can be demonstrated by inter-site comparative studies. Except where damage levels are very high, only minimal additional interventions should be required, but these will be highly specific. Where active conservation management is to be applied as part of forest policy, long-term monitoring will need to be introduced and field staff will require training in the interpretation of ecological data and the application of site-specific interventions where required.

# 8

---

# Intervention to maintain biodiversity

### The role of interventions

Primary tropical forests are characterized by a large variation in age structure and usually a high tree species richness, giving a maximum diversification of niches. There are some exceptions, however. Seral stages are more diverse than the climax vegetation in, for example, Queensland rain forests, the *Cynometra* forests of Uganda and the *Gilbertiodendron* forests in eastern Zaire.

Managed forests are composed of patches logged at different times and with a high proportion of pioneers or uniformly aged maturing trees corresponding to the time of logging. Managed forests may contain fewer large trees, a lesser abundance of standing or fallen dead wood, and a higher percentage of gaps or building phase vegetation in old gaps.

Principles of island biogeography theory have been used to design management systems for temperate forests which maximize the biological value of the management units. In effect, management attempts to match the spatial heterogeneity of natural forest. These systems revolve around the creation of a mosaic of differently aged forest patches through consecutive cutting of widely separated coupes, and the linking of patches of remnant undisturbed forest by narrow unlogged corridors (Harris 1984). This type of management has been applied fairly successfully in clear-cutting of some temperate forests and subtropical forests of southern Australia (Recher *et al.* 1987). However, it is often disliked by forest authorities because of the extra cost of creating a permanent road network and becomes redundant where replanting is preferred to natural regeneration (Franklin & Forman 1987).

Many principles carry over into the management of tropical rain forests, such as separation of logging blocks to minimize risks of spreading fires, and the requirements for refuge areas and dispersion corridors. However, the economics of road construction and maintenance, and of transport operations in most tropical forests generally preclude mosaic cutting. Logging in tropical forest typically involves the cutting of selected trees rather than clear cutting and is consequently much more extensive. Logging takes place on an advancing front, temporary roads being constructed and abandoned as logging moves into the forest block. Perhaps the most important universal principle is the need for forward planning in the design of management units. The maximization of biological value should be an element in concession planning rather than a consideration during the restoration of logged-over forests.

There is a frequent need for management intervention to promote the retention of biodiversity, even in totally protected areas where 'benign neglect' can result in eventual local extinction of species. Whether or not intervention is necessary to maintain biodiversity in production forest is broadly a question of the intensity of logging and the level of associated environmental damage. In lightly disturbed areas, such as central African forests where commercial trees are present at densities of $< 1$/ha, changes in local biodiversity resulting from responsible logging will be minimal (White 1992). In lowland dipterocarp forests, mahogany forests or other intensively logged forests, forward planning may be required if biodiversity is to be maintained.

### Habitat preservation

In many countries, measures already exist that help preserve biodiversity either as a primary or a secondary function. These generally involve the preservation of unlogged forest areas for hydrological reasons or to preserve seed trees of timber species.

One measure already commonly applied is the retention of forested strips along waterways within logging concessions. In Malaysia, for example, the law requires strips of 40 m width on each bank of rivers greater than 20 m wide, and strips of 20 m width on each bank of smaller watercourses (Munang 1987). While designed to protect water quality, these riparian strips may act both as refuges and as dispersion corridors for animal species.

Another commonly applied measure is the preservation of representative forest areas. In Malaysia there is a longstanding practice of reserving parts of forest land as virgin jungle reserves (VJRs) for the protection of seed trees of timber species, and because of watershed values, etc. Where they are not illegally logged, which is a common problem, these VJRs also maintain populations of many forest animal species (Laidlaw 1994). Similarly, it has been proposed that VJRs should occupy at least 5% of individual concession areas in Suriname (de Graaf 1986). On a wider scale, reservation of 20% or more of the total forest area in nature reserves or the equivalent has been proposed for southern Australia, Uganda and south-east Sabah (Recher *et al.* 1987; Earl 1992; Marsh & Greer 1992).

Of course, the precise percentage of forest that should be preserved is controversial. In an analysis of requirements for old-growth forest in the Pacific North-West, Harris (1984) suggested retention of 5% of the total area as a requirement for biodiversity conservation. This has been disputed by others who demonstrate that preservation of up to 20% may be required to preserve species such as the northern spotted owl *Strix occidentalis caurina* (Hunter 1990). A figure of 5–10% is probably appropriate at the level of individual concessions and the figure of 20% appropriate at the level of biogeographical regions where national parks and other protected areas come into the equation.

There is a much-debated question as to the relative merits of preserving a few large reserves or many smaller ones. At the preliminary planning stage, it is normal for the former to take priority while forests are zoned into various end-use zones. For example, the application of the simple 'Man and Biosphere' or 'Multiple-use Module' concept (Noss & Harris 1986) could be seen at the Kibale Forest in Uganda, where a central nature reserve was surrounded by buffer zone areas of non-destructive use and thus isolated from areas of intensive use (Fig. 8.1).

During more detailed enumeration within harvestable zones, smaller protected areas are commonly preserved as riparian reserves, jungle reserves or non-commercial timber stands. The large forest area managed by the Sabah Foundation in south-east Sabah, for example, allows for protection of mature forest in a number of reserve categories (Table 8.1). The location of reserved and connecting areas on the ground (Fig. 8.2) shows a typical scattered distribution of mature habitat in an intensively logged area (see also Arnold *et al.* 1993).

### Refuges

The effectiveness of a refuge in maintaining species depends on three main factors: area, distance from other refuges and the degree of difference between the vegetation of the refuge and its surrounding matrix.

The minimum size requirement of ecosystems has been much explored, both theoretically (Simberloff 1986) and experimentally (Lovejoy *et al.* 1984). Minimum sizes of refuges should ideally be calculated on the basis of home range sizes of animals (or population density, which takes into account lacunae in the distribution of individuals) and numbers required to maintain genetic diversity (stable breeding population size). Home range sizes are unknown for most tropical forest species and can be variable according to habitat type, but

Fig. 8.1. An example of forest end-use zoning, Kibale Forest, Uganda. Zones are: (1) strict nature reserve; (2) buffer zone, including (2a) ecotourism site and (2b) scientific research area; (3) silvicultural zone, including (3a) community use areas. This represents the management plan for the forest drawn up in 1992, it was altered in part following regazetting of the forest as a National Park in 1994. (Source: after Earl 1992.)

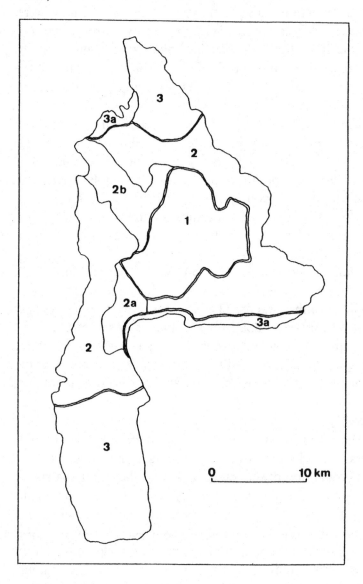

population densities reported for some large-bodied species can be extremely low.

The numbers of individuals required to make up a stable breeding population has been variously estimated at 50–1000 for animals (Simberloff 1986; Soulé 1987) and 200–500 for trees (Ashton 1976; Moran & Hopper 1987). The lower figures correspond to numbers required to withstand short-term perturbation, and larger figures to numbers required to ensure survival in the long term. In Malaysia, the area required to support minimum viable populations of 50 rhinoceros hornbills *Buceros rhinoceros* and 50 helmeted hornbills *Rhinoplax vigil* has been suggested to be at least 100 km$^2$ (Medway & Wells 1971), although this is based on atypically low population densities for these species (Johns 1987). Clearly, the provision of refuges of this size involves national legislation above the level of individual forest management units. Smaller refuges will support a proportion of resident species according to their size.

Species-area curves for refuges are logarithmic; doubling the size of small refuges will allow the retention of many more species than doubling the size of large refuges. This correlates with the number of subhabitats included in the refuge (Simberloff 1986). For example, the numbers of birds persisting in small refuges in the Kibale Forest of Uganda has been correlated with vegetation structural diversity (Dranzoa 1995). In general, wide-ranging species persist

Table 8.1. *Categories of reserved forest within a productive block of south-east Sabah managed by the Sabah Foundation*

| Category of forest | Mean size of unit (ha) | Area (ha) | % total |
|---|---|---|---|
| Nature reserves | 41 000 | 82 800 | 8.26 |
| Water catchment areas | 1630 | 8140 | 0.81 |
| Unworkable areas (steep or non-commercial stands) | 240 | 97 280 | 9.71 |
| Virgin jungle reserves (established and proposed) | 70 | 28 581 | 2.85 |
| Riparian reserves | — | 4000 | 0.40 |
| Roadside reserves | — | 500 | 0.05 |
| Total unlogged forest area | | 221 301 | 22.08 |
| Total area of block (includes 972 804 ha managed by SF and 29 466 ha of state-owned protected area) | | 1 002 270 | 100.00 |

*Sources*: Marsh & Greer (1992); A. Grieser Johns (unpublished data).

poorly in isolated ecosystems (Willis 1979). Studies in the Neotropics and peninsular Malaysia have demonstrated that large-bodied area-demanding species are lost from reserves of < 10 000 ha (Terborgh 1992, Laidlaw 1994). Hopkins & Saunders (1987) consider that in Australia any reserve of < 500 000 ha will lose some such species.

In isolated ecosystems, the long-term survival of animals will depend on the

Fig. 8.2. Distribution of reserved areas in a dipterocarp forest logging concession, Ulu Segama Forest Reserve, Sabah. Major rivers (indicated) are bordered by unlogged strips 40 m wide. Light shading represents deleted (non-commercial) areas; medium shading represents Virgin Jungle Reserves; heavy shading represent (1) a catchment area around the Danum Valley Field Studies Centre and (2) the Danum Valley Conservation Area. The map excludes small patches of forest left undamaged during logging although not specifically marked for retention. (Source: maps provided by Pacific Hardwoods Sendirian Berhad, Lahad Datu, Sabah.)

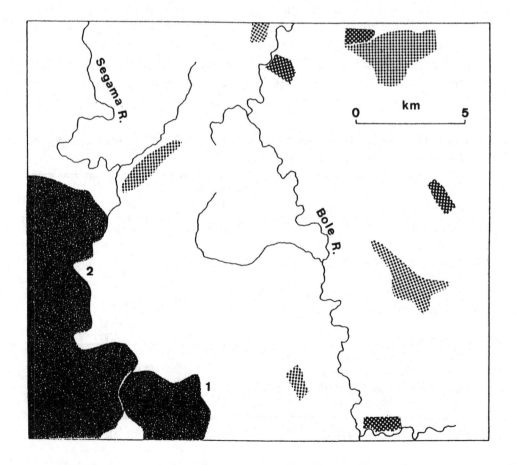

populations in the fragments being sufficiently large to remain viable, or being functionally linked so as to form a viable demographic unit. There is a fundamental difference between isolated ecosystems and refuges in production forest estate. In the latter case, differences between the mature vegetation in the refuge and the intervening matrix are less extreme; in later regeneration stages differences may be quite subtle. A refuge in a logged forest matrix will be most important at early stages of regeneration of the matrix, and may contain animals which include the matrix within their home range. A mature temperate forest refuge surrounded by late stage regeneration may contain as many species as an isolated forest patch 10 times its size (Harris 1984). Species eliminated from small isolated refuges are frequently able to persist in large areas of regenerating forest (Johns 1985; Kavanagh & Bamkin 1995).

Given a low rate of extinction from small refuges surrounded by successfully regenerating forest matrix, it is often desirable to maintain several smaller refuges rather than one large one. This maximizes the conservation of local diversity, guards against ecological catastrophes, such as fire, and aids the local dispersal of species back into the matrix (particularly in the case of trees whose fruits do not disperse widely: Hawthorne 1993). However, in determining the effective minimum area of a refuge, other factors need to be taken into account. Primary among these is edge effect, whereby the edge of the reserve may take on characteristics of the surrounding degraded matrix.

Wind penetration into isolated mature forest refuges causes rapid change in the microclimate and vegetation (Kapos 1989). In a circular forest patch of 80 ha, for example, around 75% would be affected by microclimatic change and only 25% in the centre of the refuge would retain the characteristics of mature forest. A refuge needs to be >2850 ha before microclimatic change affects <10% of its area. This has been a powerful argument for retaining large refuges. However, the above applies only to isolated refuges. Wind penetration into refuges surrounded by a logged forest matrix varies according to the extent to which the matrix vegetation is opened up during logging, and becomes minimal in later regeneration stages. Small refuges are likely to be ecologically stable in logging areas, particularly where logging is causing the loss of 50% of trees or less.

The patchwork logging systems advocated for various temperate forests, whereby refuges are always connected with unlogged forest blocks or late-stage regeneration, are unlikely to be viable in tropical forests. There will therefore be a period when the vegetation of the surrounding matrix is biologically depauperate, unless specific measures are taken to conserve matrix quality (as has been practised in the Pacific North-West in order to conserve the northern spotted owl: Thomas *et al.* 1990). Refuge areas need to be large enough to support species that are unable to persist in the matrix during this critical period. However, since most such species are small with small home ranges, this need not create an excessive demand for forest retention.

The length of the critical period will be dependent upon the efficiency of regeneration of the matrix vegetation. In degenerating logged[23:? < 50] areas of the Kibale Forest, Uganda, where basal area was only 40% that of adjacent unlogged forest, the abundances of birds within refuges remained very different to their abundance in the intervening matrix (Table 8.2). The degree of difference is likely to be less where the matrix vegetation is regenerating successfully.

Table 8.2. *The occurrence of selected understorey bird species in small forest refuges and intervening matrix at Kibale Forest, Uganda*

| Species | Numbers of individuals captured | |
| | 0.7 to 12.0 ha forest patches | Matrix |
|---|---|---|
| *Turtur tympanistra* | 12 | 12 |
| *Trichastoma albipectus* | 28*** | 6 |
| *T. fulvescens* | 8 | 17 |
| *Andropadus curvirostris* | 26 | 24 |
| *A. gracilirostris* | 41 | 51 |
| *A. latirostris* | 212* | 173 |
| *A. virens* | 18 | 33* |
| *Bleda syndactyla* | 46* | 23 |
| *Alethe poliocephala* | 98*** | 36 |
| *Cossypha cyanocampter* | 21 | 33 |
| *Sheppardia aequatorialis* | 28** | 10 |
| *Apalis binotata* | 1 | 21** |
| *Bathmocercus cerviniventer* | 32 | 48** |
| *Hylia prasina* | 10 | 13 |
| *Prinia bairdii* | 1 | 22*** |
| *P. leucopogon* | 2 | 18*** |
| *Camaroptera chloronota* | 24* | 11 |
| *Platysteira blissetti* | 12 | 10 |
| *Nectarinia olivacea* | 90 | 79 |
| *Spermophaga ruficapilla* | 8 | 27*** |

Results are based on equal mist-netting samples in the different sites. Species listed are those where $>20$ individuals were trapped.

The matrix and forest patches are located within a compartment logged 23 years previously with an unknown damage level (but probably $<50\%$).

Significant differences ($\chi^2$ test) are indicated: $P < 0.05$, **$P < 0.01$, ***$P < 0.001$.

*Source*: Dranzoa (1995).

Finally, retention of small refuges on an annual coupe basis, and the presence of linear riparian strips, means that inter-refuge distances may be relatively small. Small understorey birds can be capable of traversing long distances where necessary, providing they are not restricted by microclimatic conditions in the matrix vegetation. For example, a mustached babbler *Malacopteron magnirostre* is recorded as travelling >3 km in hill dipterocarp forest in Sarawak (Wells *et al.* 1978); yellow-whiskered greenbuls *Andropadus latirostris*, brown-chested alethes *Alethe poliocephala* and a rufous flycatcher-thrush *Stizorhina fraseri* travelled 3–5 km in medium-altitude forest in Uganda (Dranzoa 1990, 1995).

In the illustrated section of the Ulu Segama Forest Reserve, Sabah (Fig. 8.2) the mean distance between retained refuges, including corridors, was 1.2 km, with a maximum of 2.8 km. Given that the regenerating forest matrix will itself support many species, erosion of genetic diversity through low connectivity of refuges is unlikely to be problematic. Some species may not be able to travel through early regenerating matrix for physiological reasons. Providing forest regeneration is successful, however, connectivity should be achieved by about 10 years after the felling event (corresponding to regeneration of canopy cover) even for these species.

### Corridors

Forested strips are often maintained along watercourses in logging areas, principally for hydrological reasons. The forest strips filter the through-flow of water, reducing the deposition of sediment into the watercourse and reducing the leaching of nutrients (Bruijnzeel 1992). In Australian timber production forests, strips are also retained primarily for the conservation of animal populations (Lindenmayer *et al.* 1993).

The conservation functions of corridors are first, to act as additional refuge area supporting resident populations of animals, and second, to facilitate the movement of animals between areas of suitable habitat. A key role of retained forest strips, including those retained primarily for hydrological reasons, is to connect and link patches of mature forest that are not harvested for timber.

### Corridors as habitat

The North Westland Wildlife Corridor connecting areas of unlogged temperate rain forest in New Zealand has a total area of 50 000 ha, is not less than 4 km wide and effectively links populations of forest interior birds (O'Donnell 1991). Most corridors are of more modest size. In Western Australia, 74% of bird species occurred in roadside corridors 5–50 m wide, although this excluded most forest interior species (Lynch & Saunders 1991). In general terms, the width of a strip will influence its suitability as habitat. Due to edge effect, only strips > 100 m wide would be expected to retain some interior forest habitat. The vegetational composition of corridors, occurrence of gaps, etc., are all important. Narrow corridors are most effective where they contain a well-developed shrub understorey.

There have been relatively few studies of the extent to which animals persist in retained corridors in logging areas. Most of the available information concerns Australian subtropical forests (Saunders & Hobbs 1991, Lindenmayer *et al.* 1993). Arboreal marsupials showed various abilities to persist in the retained strips dependent primarily on the abundance of specific subhabitats containing specific resources. The position of strips in the landscape was an important factor in their value as temporary refuges: strips that connected ridgetops, midslopes and valley bottoms contained a greater diversity of subhabitats and supported more species. The glider *Petauroides volans* and the possum *Trichosurus caninus*, which occur singly or in pairs and are folivorous, persisted better than Leadbeater's possum *Gymnobelideus leadbeateri*, which live in colonies, have a varied diet and possess a complex social organization (Recher *et al.* 1987). The conservation of Leadbeater's possum has been actively pursued by the retention of corridors up to 200 m wide in logged[−:100] forests, but the availability of certain resources, such as tree cavities, is a better determinant of persistence than corridor width *per se* (Lindenmayer & Nix 1993). The area of habitat retained within the corridor is important in providing a minimum number of cavities and other resources (Fig. 8.3).

In Sabah, proboscis monkeys *Nasalis larvatus* which occupy linear home ranges based on the distribution of riparian vegetation, may persist in riparian reserves in logged[6−12:57] forest at largely unchanged densities (Johns 1989a). A 2 × 40-m strip width appears adequate in this case, and may be so for most species specializing in riparian habitat. In Queensland, 2 × 10-m strips of streamside habitat retained in logged[0−1:23−16] rain forest have been shown to be important not just for specialist streamside birds but for many other species that commute to the streams daily or in times of drought (Crome 1991). All bird species persisted in forests where strips of this size were maintained and the surrounding matrix was well managed.

Although many species may persist in corridors in the short term, there is evidence that some species may become more prone to predation. There may also be a higher incidence of disease in corridors. Mantled howler monkeys *Alouatta palliata* in riparian strips in Costa Rica appeared more susceptible to disease than populations in remaining unlogged deciduous forests (Jones 1994), although it is not clear why this should be so.

### Corridors as dispersal routes

The functional importance of corridors in allowing dispersion through degraded landscapes has been relatively little studied (Simberloff & Cox 1987). Most available data refer to movements of vertebrate species, although there is some evidence that corridors can be important for plant seed dispersal in allowing for movements of frugivorous birds (Forman 1991).

The extent to which animal species are selective of certain habitats during dispersion varies. Forest-dwelling kangaroos *Macropus* will move through open habitats where necessary (Arnold *et al.* 1993). Rodents tend to remain within usual habitats, moving at the fringes of occupied territories (Harrison 1992). Corridor width is a significant factor affecting the dispersion on birds in western Australia (Saunders & de Reberia 1991), primarily because of changes to the habitat caused by edge effects in narrow corridors. Recher *et al.* (1987) recommend a minimum width of 50 m to allow bird dispersion in this region.

For most species, corridor length (the distance connecting large patches of retained habitat) is important and needs to be less than the dispersal distance of the species. While transience is common in understorey birds, mammals

Fig. 8.3. Predicted abundances of the mountain brushtail possum *Trichosurus caninus* in retained corridors according to numbers of cavities and corridor length, central Victoria, south-east Australia. Dotted lines give 68% confidence limits. The corridors are c. 100 m wide, but corridor width is not a significant factor in explaining variation in possum abundance. For each prediction, values of other significant variables are kept constant. (Source: Lindenmayer *et al.* 1993.)

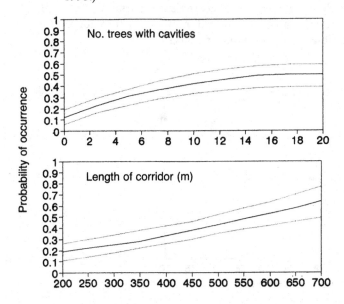

typically disperse less than five home range diameters (Chepko-Sade & Halpin 1987). Corridors with a length greater than five home range widths will need to contain suitable habitat for the species concerned if they are to assist its dispersion (Harrison 1992).

Species possess differential abilities to disperse through matrix vegetation (Diamond *et al.* 1987; Laurance 1991), particularly in logged forest, where the matrix converges with the vegetation of the corridor as it ages. Corridors may be important as dispersal routes only during the first few years after logging. They may allow the more rapid colonization of matrix vegetation by providing increased access to the matrix during early regeneration stages.

### An example of the importance of habitat preservation

In logged[6-12:32-58] dipterocarp forests of south-east Sabah, bird species-richness was comparable between unlogged forest and a 3000-ha refuge but was significantly less in a 300-ha refuge (Fig. 8.4). The numbers of species in riparian corridors, both those connected to refuges and those isolated from them, were not significantly different to those in large unlogged forest areas. This was partly due to their containing species typical of disturbed riverine vegetation, however. Some forest interior species were either absent from corridors or occurred at very low densities and were not included in the sample.

Most bird species identified as vulnerable to the effects of logging, and thus

Fig. 8.4. Cumulative bird species richness in unlogged forest refuges and riparian corridors, Ulu Segama and Tabin Forest Reserves, Sabah.
Areas surveyed are: closed rectangles, unlogged forest; cross, a 3000 ha refuge; star, a 300 ha refuge; open rectangle, a riparian corridor adjacent to unlogged forest; x, a riparian corridor surrounded by logged forest. (Source: A. Grieser Johns unpublished data.)

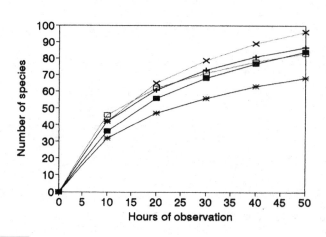

absent from recently logged forest, persisted in reduced densities in a combination of small refuges and corridors (Table 8.3). Dispersion back into the matrix appears to take place from these retained areas. There was no significant difference in the abundance of the vulnerable taxa in logged forest adjacent to or distant from a large (3000 ha) unlogged forest block (Table 8.4). This suggests that colonists originate from small refuges and riparian corridors spread throughout the logged forest rather than spreading outwards from the large forest block.

In this example, the retained areas typically occupied only 4–5% of each annual coupe (Grieser Johns 1996). Retention of this small percentage of annual coupes appeared, in this case, to allow the persistence and recolonization of most species intolerant of conditions in the recently logged forest.

### Conservation of specific resources

In addition to general habitat preservation measures, interventions may be directed towards conserving specific resources with a high potential importance to local biodiversity. Such measures are applicable within harvesting zones rather than by excluding areas from harvesting.

The most important factors affecting biodiversity are habitat heterogeneity and structural diversity. However, the maintenance of subhabitats is too general a concept to be applied in forest management. It is necessary to pinpoint specific components of diversity which are adversely affected by current logging operations and which may be manipulated within the economic limitations imposed. Components already identified as important are the density of keystone food resources, the density of deadwood and the availability of tree cavities as refuges and breeding sites. The retention of such resources can be addressed by appropriate action during inventory, felling and log transport operations (see Chapter 9).

#### Keystone food trees

Keystone food trees ('keystone mutualists') are those species which provide crucial resources for animal species at certain times. Examples would be the canopy tree *Casearia corymbosa*, which provides a resource for up to 21 obligate frugivores during the 2–6-week annual period of fruit scarcity in Costa Rica (Gilbert 1980) and *Quararibea cordata* and *Erythrina verna* which provide a nectar source used by many mammals and birds during the 8-week dry season in Peruvian Amazonia (Terborgh 1983). In a detailed study of vertebrate frugivory in south-east Bornean dipterocarp forest, Leighton & Leighton (1983) identified co-relationships between various fruiting plants and animals at times of general fruit shortage. Woody climbers of the Annonaceae and

Table 8.3. *Abundances of bird species in refuges and riparian corridors at Ulu Segama and Tabin Forest Reserves, south-east Sabah*

| | Continuous | Unlogged forest sites | | | | 6-year-old logged forest adjacent to isolated strip |
| --- | --- | --- | --- | --- | --- | --- |
| | | 30-km² refuge | 3-km² refuge | Non-isolated riparian strip | Isolated riparian strip | |
| Transect distance | 104 | 77 | 11 | 34 | 26 | 59 |
| Observation time (hours) | 94 | 90 | 16 | 35 | 36 | 46 |
| No. species accumulated in 50 hours of observation | 83 | 100 | 69 | 84 | 97 | 117 |
| $r^2$ value for logarithmic mode | 1.00 | | 0.99 | 0.96 | 0.98 | 0.98 |
| Sighting rate of vulnerable taxa[a] | | 1.18 | 2.17 | 0.31 | 0.52 | 0.29 |
| Comparison with primary forest | | 0.344 | 0.219 | −0.219 | −0.078 | −0.031 |

The taxa considered here are given in Table 8.4. Comparison statistics are *P* values, Wilcoxon signed-ranks test; a negative number indicates a reduction in sighting rate. The non-isolated riparian corridor backed onto unlogged forest, the isolated riparian corridor was approximately 40-m wide and backed onto 6-year-old logged forest.

[a]Sighting rate is proportional to the number of observations per hour in primary forest.

*Sources*: Grieser Johns (1996, unpublished data).

lipid-rich capsular fruits of the Meliaceae and Myristicaceae were considered particularly important to primates and hornbills, respectively.

Figs *Ficus* are commonly identified as a keystone resource in most geographical regions (excepting central Africa: Gautier-Hion & Michaloud 1989) since the fruit of most species are exploited by many animals. Furthermore, some species fruit several times annually, providing fruit at times of general food scarcity (Lambert & Marshall 1991). While some species are fig specialists, such as green pigeons *Treron*, the role of figs as keystones is in many cases arguable since they are generally exploited opportunistically. Many frugivorous species eat figs when available but the presence of figs is not crucial to their survival at certain times of the year. The abundance of animal species has only rarely been correlated with fig abundance (e.g. redtail monkeys *Cercopithecus ascanius* in the Kibale Forest of Uganda: Skorupa 1986; lesser mousedeer *Tragulus napu* in the Ulu Segama Forest Reserve of Sabah: Heydon 1994). Nevertheless, the presence of fig trees attracts wide-ranging frugivores

Table 8.4. *A comparison of sighting rates of understorey bird species in logged forests adjacent to and distant from a 3000-ha unlogged forest block at Ulu Segama Forest Reserve, Sabah*

| Groups of understorey birds vulnerable to logging | Sighting rate[a] | | | |
| --- | --- | --- | --- | --- |
| | 6-year-old logged (adjacent) | 6-year-old logged (distant) | 12-year-old logged (adjacent) | 12-year-old logged (distant) |
| Phasianidae[b] | 0.25 | 0.10 | 0.17 | 0 |
| Trogonidae | 0.63 | 0.47 | 0.81 | 0.42 |
| Turdidae[c] | 0.16 | 0.20 | 0.34 | 0.81 |
| Terrestrial Timaliidae | 0.27 | 0.99 | 1.20 | 0.26 |
| Muscicapidae[d] | 0.45 | 1.24 | 1.62 | 0.52 |
| Comparison with primary forest | −0.031 | −0.078 | −0.156 | −0.016 |
| Inter-site comparison | 0.406 | | −0.156 | |

Comparison statistics are *P* values, Wilcoxon signed-ranks test; a negative number indicates a reduction in sighting rate.

[a]Sighting rate is proportional to number of observations per hour in primary forest.

Known secondary species are excluded: [b]excluding *Coturnix chinensis*; [c]excluding *Copsychus saularis*; [d]excluding *Hypothymis azurea* and *Rhipidura javanica*.

*Source*: Grieser Johns (1996).

which may defecate or regurgitate seeds from elsewhere and so assist forest regeneration.

Because of the canopy germination habit of most figs, protection of existing mature individuals is the only viable measure of conservation during felling exercises. The density of fig trees is generally less in recently logged forest, relative to the extent of damage (e.g. a reduction from 3.3 to 1.4/ha in logged[2:?60] forest in south-east Borneo: Leighton & Leighton 1983). Regeneration of figs usually occurs and the abundance of figs in older logged forests can be greater than that of unlogged forest (Fig. 8.5), but this varies between forests. In some cases, greater fig abundances reported in older logged forests are due primarily to vigorous regeneration of small free-standing fig species rather than recovery of numbers of strangling species (Heydon 1994).

Definition of true keystone food resources is problematic in that the level of specialization of feeding behaviour in many tropical species is high. Most tropical trees have specialized insects feeding on them, and thus each tree is a keystone resource to a certain group of taxa. Even in temperate forests, frugivorous birds can be highly specific, as in the case of crossbills *Loxia*: each of 5–7 sympatric species in North American coniferous forests specialize on a single or a few conifer species (Benkman 1993). Selection of a keystone tree

Fig. 8.5. Fig tree density in managed mahogany forest, Budongo Forest Reserve, Uganda. Forest types are as follows: (1) climax *Cynometra* forest; (2) climax *Cynometra*/mixed forest; (3) logged *Cynometra*/mixed forest; (4) climax mixed forest; (5) logged mixed forest. (source: adapted from Earl 1992.)

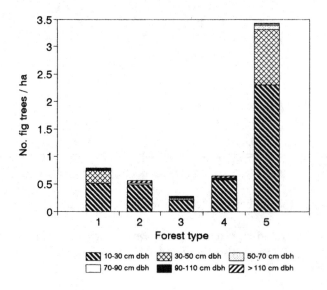

species for management purposes broadly implies the selection of certain taxa for conservation attention. This may have merit in the conservation of animal species known to survive poorly in logged forest, but should not take precedence over overall reduced-impact logging where the entire range of food resources will be less adversely affected.

### Standing and fallen deadwood

Standing dead trees ('snags') may make up 5% of the trees in a forest and are an important structural element in maintaining the diversity of habitats. They create an opening in the canopy that allows light penetration, which causes vegetation growth throughout the vertical profile rather than simply on the ground (as is the case in gaps). This is particularly important for epiphytes. They also provide deadwood substrata for various wood-feeding insects, which form a food supply for a variety of other animals. Finally, they are important for animals which use tree cavities as refuges or breeding sites. Different suites of animals are associated with dead trees at different levels of decay (Lindenmayer *et al.* 1991).

In unlogged tropical forests, fallen trees and logs can reach densities as high as 600 tonnes/ha, forming a further important component of habitat diversity. The deadwood itself is used by termites and various other insects as a food source, provides refuges for amphibian, reptile and small mammal species, and provides shelter for ground-nesting birds. A loss of the deadwood microenvironment can have serious consequences for associated animals (Du Plessis 1995).

In a Douglas fir ecosystem of the Pacific North-West, 45 vertebrate species were shown to be reliant on standing and fallen deadwood (Harris 1984). The presence of deadwood was found to be the single most important factor affecting differences in vertebrate populations between mature forest and managed forest. The number of vertebrate species decreased by 10% where standing deadwood was absent and by 29% where both standing and fallen deadwood was absent. Similarly, the abundance of deadwood has been correlated with the density of cavity nesting birds in temperate forests of western Oregon (Schreiber & deCalesta 1992), Swedish pine forest (Sandström 1992) and temperate riverine forest in South Africa (Du Plessis 1995). Possible co-variance in the abundance of deadwood and animal groups has not yet been addressed in studies in any tropical forest.

The number of snags required to support dependent species has been predicted for a North American temperate forest based on ecological requirements of woodpeckers (J. W. Thomas, in Hunter 1990). This requires information on the maximum population density of each woodpecker species, the average number of snags used annually by each pair and the number of snags not used. In this case, maximum populations of five sympatric woodpecker species could be conserved by retaining $0.7 \times 25$ cm diameter snags, $3.4 \times 30$ cm diameter snags and $0.3 \times 50$ cm diameter snags/ha. In Swedish pine forest, it was calculated that $>14$ snags/ha between 28 and 128 cm diameter and between 6.4–25 m height were required to maintain populations of all associated birds (Sandström 1992). This implied a need both for existing snags and moribund but still living trees $>70$ years old to be retained to maintain the cycle of snag development.

At Sungai Matan, Indonesia, the density of large dead trees was found to be higher in older logged[8:?70] forest than in nearby unlogged forest (Cannon *et al.* 1994). This was not the case in recently logged[1:?70] forest, suggesting that many old trees damaged by logging eventually died, supplementing the deadwood resource. Under heavy logging levels the deadwood resource may not be limiting in early regeneration stages but may become so as trees killed by logging disintegrate and numbers of new snags are very low. Under light logging levels, the natural cycle of snag formation is likely to be less affected providing that dead or overmature trees are not targeted by timber stand improvement practices. If snags are critically reduced during logging, deadwood can of course be created by killing selected trees (Hunter 1990). However, it can be expensive to create naturally decaying snags and careful criteria are required to select the individual trees so treated.

### Tree cavities

Tree cavities are used by a large number of tropical forest animals for refuges and breeding sites. They may be formed naturally as part of the growth form of certain trees or by rotting out of wounds or heartwood, or they may be created by chiselling out of live wood by certain bird species.

Species dependent on cavities either for breeding or as refuges or sleeping sites include a variety of large birds (e.g. hornbills, owls, parrots, barbets), squirrels, small primates (such as lion tamarins *Leontopithecus* and many prosimians), marsupials and bats. In Australian forests, where there are no species capable of excavating their own cavities, more than 350 vertebrate species are dependent on naturally forming cavities as nesting or refuge sites (Ambrose 1982).

Cavities used by large species, such as hornbills or owls, generally result only from the rotting out of broken branches or boles. Forest stands less than 100–120 years old do not contain significant amounts of heart rot and generally do not develop the large cavities used by such species (Harris 1984; Lindenmayer *et al.* 1990a, 1993). However, damage to residual trees during logging can result in new cavities being formed, although they may take some time to rot out.

## Cavities as limiting factors

Clearly, the availability of cavities is most likely to become limiting in forests that are clear-felled. In montane ash forests of southern Australia, where cavities in unlogged forest have a high (36%) occupancy rate, clear-felling on an 80 to 120-year cycle effectively eliminates cavity-dwelling species in the absence of remedial measures (Lindenmayer *et al.* 1990a). A few cavity-dwellers are able to occupy alternative refuges or breeding sites, but these are rather an exception. In the Atlantic forests of Brazil, the golden lion tamarins *Leontopithecus rosalia* normally use cavities as sleeping sites. However, the species has been recorded living entirely within 10 to 15-year-old scrub regenerating on cleared land. No cavities were available in this regenerating vegetation, but the lion tamarins were able to use liana tangles or other refuges instead (C. Peres, pers. comm.).

The availability of cavities may also become limiting in even-aged timber stands developing subsequent to very high logging levels. There are likely to be fewer large old trees in these regenerating forests and fewer large cavities. Small cavities may develop in quite young trees, and certainly in trees that are still living, but the size of available cavities is important to some animal species. The abundance of excavated cavities may also be reduced in even-aged timber stands since those species which excavate their own nest holes tend to use snags in preference to living trees. In temperate forests of South Africa, removal of standing deadwood for fuel has caused a significant reduction in the density of excavated cavities, but did not always affect the density of naturally forming cavities (Du Plessis 1995).

In less intensively managed forests, the availability of cavities is less likely to become limiting. In a degenerating logged[23:? < 50] compartment of the Kibale Forest of Uganda, Dranzoa (1995) recorded the density of naturally formed cavities as 26/ha, compared with 62/ha in adjacent unlogged forest. In this example, 51% of cavities in unlogged forest and 43% of cavities in the logged forest showed evidence of occupation by birds, although the number used per breeding season was less. This is a higher occupation rate than has been recorded in temperate forest. Sandström (1992) found that only about 7% of available cavities were occupied by nesting birds in both unlogged and logged pine forest. In neither of these forests was the availability of cavities considered to be a limiting factor. The density of breeding pairs of cavity nesting birds was less in the logged forest in both examples, but this correlated more with features of environment and food supply.

Populations of large hole-nesting birds, such as South-East Asian hornbills, do not appear to be depressed in forests logged at moderate intensity (see Chapter 5). The density of breeding sites of black-and-white casqued hornbills *Bycanistes subcylindricus* in degenerating logged[13:?<50] forest at the Kibale Forest, Uganda, has been reported to be less than in adjacent unlogged forest (Kalina 1988), but very small sample sizes were involved. Hornbill abundance has nowhere been correlated with the abundance of large limb cavities.

### Primary hole nesters

Relatively few species excavate their own cavities (primary hole nesters). Almost all of these are woodpeckers (Picidae). Woodpeckers have been suggested to be keystone species, excavating cavities which are subsequently used by other species (secondary hole-nesters). Woodpeckers have been reported to decrease in density in logged[8:57] forest in Sabah (Lambert 1992), although a previous study at the same logged[6:57] forest site did not document significant differences (Grieser Johns 1996). Where they occur, reductions in woodpecker density are likely to be due primarily to a loss of foraging sites, but may well be reflected in a drop in the numbers of excavated cavities available to secondary hole nesters.

### Creating or retaining cavities

In some temperate and subtropical forests where the availability of nesting sites has been shown to limit the density of cavity-dwelling species, provision of nest-boxes can help maintain their populations (Menkhorst 1984; Newton 1994). This is not likely to be an option in most tropical forests because of the expense and logistic difficulties involved, and because in many areas the contents of the nest boxes would be viewed as an easy source of protein by local people (Du Plessis 1995).

Within southern Australian forest clear-felled on an 80 to 120-year rotation, legislation has been applied to retain snags specifically for cavity-dwelling animals. For example, rules applied in the Dandenong region require 15 snags to be left per 10 ha of clear-felled forest (Lindenmayer *et al.* 1990a). This is rather a minimal provision since more than one species of marsupial occupies the same tree in <1% of cases, and this would allow for retention of only a very low marsupial density. Furthermore, isolated standing trees tend to suffer high mortality through windthrow or fire (Lindenmayer *et al.* 1990b). Retention of trees in clumps has been suggested as an alternative solution, although in this particular case a clumped spatial distribution of trees does not appear to promote retention of larger numbers of marsupials.

Within selectively logged forest, particularly those incurring damage levels of < 50%, active retention of snags is less likely to be necessary. Many factors determine whether cavities are occupied or not, however, such as the degree of isolation of the tree and the habitat surrounding it. If occupancy rates are less in logged forest, due to a smaller proportion of trees remaining suitable for cavity-dwellers, then corrective measures may be required.

### Enrichment planting

Replanting of commercial species has been a widespread means of supplementing inadequate natural regeneration. Seedlings are raised for the purpose, usually of indigenous pioneers with a minimum growth rate of 1.5 m/year. The seedlings are planted in natural gaps, along regularly placed lines, or on heavily damaged open areas such as roads and loading areas. Planted-out seedlings usually need to be tended over a number of subsequent years, which in dipterocarp forest can increase the timber production costs per $m^3$ by 4–40 times compared with natural regeneration systems. The success of establishment and commercial increment of the plantings has been mixed (HIID 1988; ITTO 1989).

Enrichment planting is normally applied in heavily damaged logged-over forest where the upper canopy has been largely removed during felling or through post-felling arboricidal treatment. In such cases, replanting may be necessary to assure any sort of second crop and prevent the excessive growth of climbers. However, the presence of large browsing mammals in the forest is incompatible with replanting of palatable species, which limits what can be planted without the need for control measures. The conflict inherent in promoting forest elephant populations is an interesting example (e.g. Hawthorne 1993). Elephants can be an important dispersal agent for the seeds of certain trees. On the other hand they may selectively browse certain tree seedlings and can in extreme cases prevent the regeneration of natural forest, as in certain compartments of the Kibale Forest in Uganda.

Suggestions have been made that replanting could include species other than commercial timbers: keystone food trees, for example. Group planting of a mix of commercial and keystone trees has lately been proposed in the restoration of degraded forest in Uganda (D. Earl, pers. comm.) and following intensive logging operations in Sabah (P. H. Moura-Costa, in Heydon 1994). Promotion of keystone trees as well as commercial trees during refining exercises was at one time proposed for dipterocarp forest in Sarawak (Proud & Hutchinson 1980), but refining as a management tool requires extremely careful use to prevent the loss of forest diversity. There are no examples of the success of mixed replanting in improving the habitat quality of regenerating forest.

During the 1960s, areas of the West Mengo Forest Reserve, Uganda, which had been heavily damaged by logging[25-30:>50] and charcoal burning were replanted with selected timber species, including exotic hardwoods. This resulted in a considerable improvement in timber increment, but led to a 17% reduction in tree species diversity compared with forest that was allowed to regenerate naturally (Dranzoa & Johns 1992). However, the bird and mammal communities of the enriched and non-enriched forests were not significantly different. This may be partly due to planting of the exotic *Maesopsis eminii*. While not a keystone food tree, this species provided a major fruit source for a variety of frugivorous mammals and birds and its associated folivorous insects provided additional resources for foliage-gleaning bird species.

Dominance of managed forests by a few commercial species of even age would be expected to cause gradual losses of animal species commensurate with a high loss of vegetation structural diversity. The number of logging cycles required to achieve a uniform stand in tropical forest is likely to be high, but related to felling intensity. If management of tropical forests concentrates on replanting of a few species to the detriment of tree diversity, over several logging cycles the forest will approach plantation conditions.

### Re-logging schedules

The long-term sustainability of the forest ecosystem will ultimately depend on whether or not successful regeneration is achieved during the applied logging cycle. Removal of $10 \, \text{m}^3/\text{ha}$ of timber in the Bia Forest of Ghana has been shown not to affect plant species diversity (Hawthorne 1993). In New Guinea, where $20–30 \, \text{m}^3/\text{ha}$ of timber is harvested, plant diversity has been shown to recover within the cycle proposed (R. Johns, 1992). In forests where damage levels are high, this is less assured (except in monodominant forests where diversity is naturally higher in non-climax stages). Furthermore, the presence of equal plant diversity does not necessarily imply the presence of all age classes, and thus the presence of original forest structural diversity. No precise data are yet available documenting the recovery of biodiversity over a complete logging cycle, and only one study (that in the Tekam Forest Reserve of peninsular Malaysia) has so far been established with this aim.

Regeneration of a timber crop can take place relatively quickly, perhaps after as little as 35 years under some projections. Re-logging frequently takes place after a lesser interval, but this will involve removal of newly commercial species or smaller size classes of traditional commercial species. Forest structural diversity will not be regained during short logging cycles and certain classes of organisms, such as canopy epiphytes and associated invertebrate species, may not recolonize by this time. These will therefore persist only in preserved forest areas. Full recovery of vegetation structural diversity may take more than 100 years. Selective logging cycles of this length would be economically unjustifiable as a form of land use, although this rotation time may be allowable for clear-felling regimes (e.g. 80–120 years is projected for clear-felled montane ash forests in southern Australia: Lindenmayer *et al.* 1990a).

Some measure of deflection of the form of the ecosystem will be inevitable in forests managed over the long term for timber production. This will be due both to the loss or reduction of some subhabitats and to the common commercial requirement for increasing the standing timber volume. However, the extent of deflection can be ameliorated by appropriate interventions. Deflection need not extend to local extinction of species if appropriate action is taken: it should be expressed primarily in changes in species abundances. In the absence of adequate intervention, a gradual loss of species over repeated logging cycles might be expected. However, this need not occur if the forest is well managed and requirements for biodiversity retention considered as part of management techniques.

### Compensatory plantations

Plantation forestry is a comparatively recent phenomenon in the tropics, most major plantations having been established since 1950. The majority of plantings involve exotic species, notably pines and eucalypts, which have the advantage of being comparatively free of pests and diseases (Zobel *et al.* 1987). Where pests or diseases take hold, however, the consequences can be very serious (e.g. the loss of *Agathis* plantations in Queensland due to an outbreak of the coccid *Coniferococcus agathidis*: Queensland Department of Forestry 1987; the loss of cypress *Cupressus lusitanica* in east Africa due to attack by the aphid *Cinara cupressi*: FAO 1991b).

Pest and disease problems have often contributed to the poor performance of many native hardwoods grown in experimental plantations: only teak *Tectona grandis* has shown consistently good growth rates. However, there are some native hardwoods that show promising initial results, such as *Cordia alliodora* or certain *Shorea* spp. (Kanowski *et al.* 1992). Some of these timbers are already acceptable within the international market, such that problems of inertia in the timber trade (a slow uptake of new species) can be avoided.

There have been some reservations concerning the extent to which there may be surface erosion under plantations, particularly of conifers, where the development of an understorey is suppressed. In Queensland, the establishment of plantations on slopes of > 15–20° is generally avoided (Queensland Department of Forestry 1987). In the Kibale Forest of Uganda it has been suggested that leachates from pine plantations may cause acidification of ground water and subsequent mortality of natural forest trees (Struhsaker *et al.*

1989), although causal relationships have not been proved. Nonetheless, timber plantations located outside of the production forest estate may have an important potential role in lessening the pressure on natural forest resources.

### Plantations as a means of reducing deforestation

Most tropical deforestation is caused by shifting agriculture, not through effects of the timber industry directly. Thus an increased provision of plantations is not likely to reduce the levels of forest clearance. Rural populations will still require land to grow crops even if their fuelwood supplies and construction timbers are provided for from outside the forest.

A further problem arises in that if plantations become regarded as the primary sources of wood products then the perceived value of natural forest could fall, leading to increased deforestation for agricultural development (Ball 1992). The balance of production from managed natural forest and plantations needs to be balanced quite carefully to ensure that natural forests remain an important financial asset, encouraging good management practice, and that it does not become cost-effective to replace them with industrial plantations or other forms of land use.

### Plantations as substitute timber sources

The compensatory role of plantations has been regarded as somewhat limited in that they cannot provide a source of large-size sawlogs and sawn timber of the types currently forming the mainstay of the international timber trade. For this reason it has been assumed that, under current market conditions, they do not form an alternative to commercial logging. Increased provision of plantations would thus not be likely to affect the volumes of timber extracted from natural forest unless there is a change in marketing strategies.

While this argument is true at the present time, there is increasing evidence that plantations are able to produce certain types of hardwood product. A. Leslie (in Ball 1992) states that the production of veneer-quality logs from plantations in South-East Asia is sufficiently developed that it will soon supplement and even supplant the supply from natural forest. This provides some support for Poore *et al.* (1989) who suggest that timber production from natural forest should be restricted to high-value decorative, joinery and veneer species that cannot be grown in plantations at the present time: efforts should be directed towards supplying other wood products from plantations.

It is certainly true that acceptable species for many end-uses can be grown in plantations. In fact, plantations already produce 7–10% of the world's current industrial roundwood from only 2.6% of the forest area. However, inertia in the tropical timber trade coupled with the high economic returns from logging natural forest (where plantation establishment costs do not have to be factored in) means that intensive logging of primary natural forest is likely to continue. Plantation-produced timber may compete better, in economic terms, with the costs of producing a second and subsequent crops from natural forest.

The second important compensatory role of plantations is in the provision of fuelwood or construction timber for local communities. This may have a major potential impact on the conservation of forest resources, particularly where natural forests are reduced to fragments and local human population density is high. While initially requiring intensive tending, plantations can be established on otherwise unproductive land, such as *Imperata* grasslands or exhausted farmland (even if this is more difficult than establishing them on more fertile, more recently cleared forest land).

Plantations of the second kind have unfortunately not been very successful to date, primarily for social reasons and because of a lack of technical expertise (Kanowski *et al.* 1992). It is frequently suggested that it would be a more secure investment to improve the management of natural forest. However, particularly for countries whose natural forest resources are severely depleted, plantation establishment would appear to be a means of reducing the need for importation of basic wood products.

### Plantations as buffer zones

Plantations at the forest edge may have a dual role in protecting the integrity of forest protected areas, whether production forest estate or preserved forest. First, they may help stabilize the forest boundary in terms of lessening edge effects through creation of a 'soft' rather than a 'hard' edge. Second, they may provide additional habitat for forest animal species thereby increasing the area available to them.

### Reducing edge effects

Forest/agricultural interfaces typically demonstrate a pronounced edge effect. Non-forest microclimatic and vegetational characteristics can penetrate considerable distances into otherwise undisturbed forest areas. Edge effects are a particular problem for small isolated reserves or those bisected by roads or containing agricultural enclaves. In fragmented portions of the West Mengo Forest Reserve, Uganda, the bird species composition of a small unlogged forest can be seen not to differ significantly from rather larger areas of logged[25-30:>50] forest (Table 8.5). Logged compartments have lost forest interior species due to the high damage levels incurred by logging and the subsequent microclimatic and vegetational changes. The unlogged forest reserve has lost the same forest interior specialists as a result of edge effect.

Establishment of tree plantations at the forest edge can ameliorate edge effects considerably. The microclimatic conditions associated with edge habitat may be displaced from the edge of the natural forest to the edge of the plantation. Conditions typical of the forest interior can extend right up to the border of the natural forest, which, in the case of small forest isolates, can increase their effective area considerably.

### Provision of additional habitat

The extent to which tropical forest animals are able to utilize tree plantations is limited, but depends on the age of the plantation, management factors, etc. (Wilson & Johns 1982). Relatively few species may be expected to range solely within plantations. These will generally be species associated with edge habitat or early stages of gap regeneration in natural forest. A larger number of species will include plantations adjacent to natural forest within their overall ranges.

At Taliwas, Sabah, plantations of *Albizia* exist adjacent to a large block of natural forest. The number of large mammals encountered along a transect from forest interior to plantation interior does not vary significantly (Table 8.6). Frugivores, such as Bornean gibbons *Hylobates muelleri* do not occur in the plantation but deer *Muntiacus muntjac* and *Cervus unicolor*, bearded pigs *Sus barbatus* and civets Viverridae range through both habitats. On the other hand, there is a pronounced fall-off in the number of bird species from the natural forest boundary into the interior of the plantations: 36% of the natural forest species do not range into the plantations at all.

Plantations provide additional habitat for only a subset of the natural forest fauna. The number of species involved depends on the species of tree and form

Table 8.5. *Bird populations in hard-edged forest remnants, southern Uganda*

| | Forest reserve | | | |
| --- | --- | --- | --- | --- |
| | Sango Bay Group | Mpanga | Buto-Buvuma (West Mengo) | |
| Forest reserve size (km²) | 151 | 4.5 | 15.3 | |
| Management history | Partly logged | Unlogged | Logged | Logged and replanted |
| Number of bird species | c. 200 | 128 | 127 | 133 |
| Percentage of forest interior specialists[a] | 41 | 23 | 23 | 22 |
| Capture rates of forest interior specialists[a] | No data | 0.78 | 0.77 | 1.03 |

[a]Bird species considered forest interior specialists are listed by Bennun *et al.* (in press).
*Sources*: Dranzoa (1990); Howard (1991); Dranzoa & Johns (1992).

of management. For example, in plantations established adjacent to the Kibale Forest, Uganda, considerably fewer bird species occur in pine than in eucalyptus plantations (D. Pomeroy, pers. comm.). Thinned pine stands with a well-developed natural understorey support more species than poorly managed pine stands with very little natural understorey.

The possibilities for managing plantations to promote the establishment of an understorey of native species has been demonstrated elsewhere (e.g. *Arucaria cunninghamii* plantations in Australia: Queensland Department of Forestry 1987; *Pinus caribaea, P. patula* and *Cupressus lusitanica* plantations in Uganda: Chapman & Chapman 1996). In some cases, plantations may actually be an efficient intermediate stage in the restoration of natural forest on degraded land. In all cases, the development of a natural understorey can increase the value of plantations for biodiversity conservation to a considerable extent.

### Summary

The need for management intervention to conserve biodiversity depends to a large extent on the damage levels associated with the logging operation concerned. If forests are subjected to very low damage levels specific conservation intervention is probably not necessary. Under conditions of reduced-impact logging, where rules are scrupulously followed, only minimal conservation intervention may be required. However, in conditions of high intensity logging the introduction of conservation-oriented policy may be important: a number of interventions can be applied at minimal economic cost to assist the retention of local biodiversity.

The most common interventions already applied at the planning stage is the

Table 8.6. *The occurrence of mammal and bird species through a forest-tree plantation interface at Taliwas, Sabah*

| | Forest interior | Forest edge | Albizia plantation | |
| --- | --- | --- | --- | --- |
| | | | 1 km from forest edge | 2 km from forest edge |
| Number of large mammal species | 9 | 11 | 7 | 7 |
| Number of bird species | 162 | 122 | 92 | 45 |
| Number of bird species restricted to habitat | 59 | 0 | 38 | |

Sources: Duff *et al.* (1984); Mitra & Sheldon (1993).

retention of preserved areas. These include areas set aside as watersheds for conservation of tree genetic resources, riverine corridors retained for hydrological reasons and specifically designated wildlife corridors. Areas set apart as non-commercial also fall into this category. During the short term, the retention of 5–10% of forest concessions as preserved areas may retain the majority of forest species in the concession as a whole. Over the long term, the survival of biodiversity is also dependent on good management of the regenerating logged forest matrix.

Intervention at the inventory and harvesting stages can take the form of marking of specific resource trees for retention. These may be either important food resources ('keystone food trees') or snags. The latter are vital to many animal species as foraging substrata and as providers of cavities used as refuges or breeding sites. The identification of keystone food trees is difficult, probably very site-specific and implies some level of selection as to which components of overall biodiversity are preferred for active retention. Retention of minimal snag densities is a more prescriptive treatment and can be effective, particularly given a detailed knowledge of the use of snags of different decay stages by different suites of animals.

Post-harvesting intervention usually takes the form of enrichment planting to enhance timber crops. Replanting of keystone food trees could equally well be applied if identified species are shown to have been critically reduced during logging. and if they grow successfully in the logged forest environment. Regeneration of timber crops can occur rather more quickly than regeneration of forest structural diversity and associated animal species composition, such that interventions to maintain tree diversity may become more important after the first rotation. Some level of change to the ecosystem may be expected in managed forest, particularly where replanting of timber species has occurred or other methods have been applied to increase the representation of commercial tree species. In well-managed forest this need not extend to a reduction in biodiversity, however, only to changes in the relative abundance of species.

Compensatory plantations have been a neglected intervention and may assist the retention of forest biodiversity in two ways. First, they may remove some of the pressure from natural forest in providing an alternative source of certain wood products, thus lessening the extraction and associated damage levels from natural forest. Second, plantations adjacent to natural forest can reduce edge effects and provide additional habitat for certain forest animal species.

# 9

## Field procedures

### Damage limitation

Reduced-impact logging is considered essential if tropical forestry operations are to be improved (Palmer & Synnot 1992). The effects of uncontrolled logging on tree regeneration in particular, and on biodiversity and ecosystem function in general, are unnecessarily detrimental to long-term sustainability of both timber production and maintenance of the ecosystem.

In areas of intensive logging, such as Sabah, a correlation has been suggested between damage levels and the basal area of timber extracted (Nicholson 1958). While the ecological and environmental impact of logging on a regional or forest-type level is broadly related to logging intensity, there is a poor overall correlation due to different local factors (e.g. logging damage in Amazonian várzea is much lower than expected due to the use of waterborne log transport). The general correlation seen among intensive logging operations is due primarily to the fact that most pay scant attention to reducing the incidental damage associated with felling and skidding. Potential minimum damage levels, which can be calculated from a simple model, have generally not been achieved in commercial operations, with the exception of those in Queensland (Fig. 9.1).

Damage limitation is a standard management procedure in most temperate forestry operations (Bol & Beekman 1989). Where such measures have been applied experimentally to tropical forests, damage levels have been reduced by as much as 50% (Marn 1982; Jonkers 1987; Malvas 1987; Hendrison 1990). Applying such measures has been shown to reduce economic costs in the above studies, which are based on small experimental plots. Some associated costs were recorded in Queensland, where such measures were applied over large areas as part of standard management practices, but these costs were shown to be minimal (Bruijnzeel 1992).

Recent concepts of best practice or wise management of tropical forest concentrate primarily on demonstrable techniques for reducing environmental damage and thus improving forest regeneration. Damage limitation is undoubtedly the key factor influencing the conservation of biodiversity in managed forest. However, as discussed in the previous chapter, there are other field procedures that may be applied at no or minimal cost aimed specifically at increasing the biological value of the residual stand. The following sections consider standard field procedures in tropical forest management operations and the value and possibilities for incorporating new procedures at each stage.

### Guidelines for conserving biological diversity

From an understanding of basic ecological principles, it is comparatively simple to produce guidelines introducing a component of biodiversity conservation into forest planning and management. However, local conditions faced by forest managers are highly variable such that it is difficult for specific recommendations to be made without a study of the specific forest type,

Fig. 9.1. Damage levels under different logging intensities (all figures for forests <1 year post-logging). Broad geographical regions sampled are: closed rectangle, Africa; cross, South-East Asia; star, Neotropics, open rectangle, northern Australia. All points are commercial mechanized operations except (a) várzea forest where machines are not used, and (b) a reduced-impact logging demonstration plot. The power curve represents a fit ($r = 0.99$) to points generated by a minimum damage model. The model makes the following assumptions: trees to be logged are randomly distributed, skidroads follow the shortest possible route between stumps and winches are not employed, each felled tree destroys five others, and 6% of the area is destroyed by construction of main access roads, loading points, etc. (Source: after Johns 1992.)

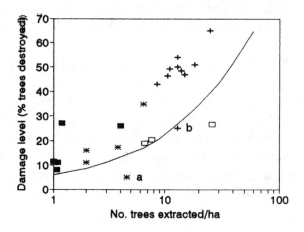

topography, or other salient features, and local socio-economic conditions. General guidelines rely heavily on local interpretation and thus on the level of training, awareness or motivation of planning and field staff.

Complementing the guidelines for sustainable management produced by the International Tropical Timber Organization (ITTO 1990), guidelines for the conservation of biological diversity in managed forest have been compiled (Blockhus *et al.* 1992). These consist of general principles concerning the potential value of managed forest for conservation and environmental services within an economic framework, and some recommended action at the level of planning and management. The principal recommended actions are:

1. Creation of a totally protected area system linked by natural or near natural forest corridors. This should be independent of the production forest estate.
2. Protection of the integrity of the forest estate.
3. Strengthening of forest authorities, both financially (through increased revenue from forest rents, etc.) and in terms of training (particularly in the sphere of biodiversity management).
4. Zoning of production forest estate according to its importance for biodiversity, with special rules applied in important forests.
5. Application of the ITTO (1990) guidelines in all production forests, with particular observance of the protection of riparian corridors and refuge areas within logging concessions.
6. Adoption of minimal intervention systems using natural regeneration of commercial and non-commercial trees where possible. Avoidance of the use of arboricides or other means of suppressing certain tree species (particularly keystone species) and retention of old and dead trees in the residual stand.
7. Establishment of ecological monitoring programmes to determine changes in biodiversity and needs for remedial action.

Recommendations 1–4 involve planning at the national or regional level and are subject to a variety of socio-economic constraints. Recommendations 5 and 6 concern the management of field operations, which will be dealt with in detail below. Recommendation 7 is a highly desirable addition to forest management practice but one which has rarely, if ever, been applied to tropical forest management and one which requires careful definition.

### Pre-felling procedures
The level of management planning in field operations varies considerably. In extreme cases, contractors may be authorized to enter a felling

block without any prior data collection. At the other end of the spectrum, rigorous pre-felling planning may be applied (an example for a forest reserve in Sabah is given by Kleine & Heuveldop 1993). In most forests, four basic procedures are applied, as follows.

### Silvicultural enumeration

Before tendering forest blocks for logging concessions, enumerations are normally carried out to assess the timber volumes available and the stocking levels of small-sized commercial species. This will determine the feasibility of different types of forestry procedures, including special harvesting methods or post-felling treatments that may be required. Enumerations are normally undertaken in sample quadrats. Often 1-ha permanent sample plots (Synnott 1979) are established at this time to provide baseline information on forest recruitment and growth.

Before concessions are demarcated and tendered, enumeration data may be compared with available data concerning areas of particular significance for biodiversity conservation (e.g. areas with a high occurrence of endemic species). Distributional data is available in many, although not all, tropical countries to differing levels of definition. In countries where GIS-based vegetational mapping and species atlases are completed, such as Kenya, the forest estate may be graded not only on the basis of timber stocks but also according to biodiversity criteria. Ideally the demarcation of concession areas and allowable offtakes and harvesting methods within them should then reflect a compromise between economic and conservation needs. In countries where detailed species distributional data are lacking, the proportion of retained unlogged forest should be high, with the possibility of down-grading when biodiversity information becomes available. The mapping of biodiversity in production forest estate, particularly in countries with high levels of endemism, should be considered a priority of the conservation community.

### Topographic survey

In most logging operations, a reconnaissance level survey will be carried out to prepare a terrain map, usually on a scale of 1:5000. This can be carried out using recent aerial photographs or by means of ground surveys. Where recent contour maps are available, it is normal for these to be used as the basis for topographic surveys. If they are not available, and forests on steep or dissected terrain are to be worked, 5- or 10-m contour intervals should be added to the survey maps.

Topographical features recorded on these initial maps are as follows:

1. Physical features (rivers, streams, swamps, rocky outcrops, etc.). These will already be marked if a pre-existing topographical map is being used as the basis for the survey.
2. Areas of non-commercial stands or areas that cannot be worked for physical reasons.
3. Areas of commercial timber stand that are to be retained (river corridors, nature reserves, etc.).
4. Annual logging coupes. Annual coupes will be decided on the basis of the above factors.
5. Logging zones or units. This will depend on condition of the timber stand, itself dependent on soils and topography. Areas that can be worked during rains should be highlighted.
6. Road and skidroad traces, indicating type of road and gradients where applicable. In areas where logging has already taken place, locations of previous roads and skidroads should be marked as a basis for a restored road system.

The physical location of riparian strips and non-workable stands is fixed. The location of reserves is more fluid and should complement the other retained areas. Access to watercourses is important for many species (Crome 1991) and reserves should always border a riparian strip. On the other hand, they should consist mainly of interfluvial or other subhabitats not otherwise retained. Criteria for the siting of reserves are discussed by Laidlaw (1994).

The total area of retained forest within a concession might be determined on the basis of overall biodiversity value at the enumeration stage (above) but should not be less than 5% and preferably 10%. The division of allowed reserve area (i.e. reserve size) could consider habitat heterogeneity and, ideally, the home range sizes of species that do not persist in recently logged forest. Unfortunately, these are rarely known. In general, concessions might aim to include one reserve of 3–5 km$^2$ per 100 km$^2$ area plus several smaller ones. Siting these close to unworkable areas might increase their effective area, and they should be well linked with the rest of the 100-km$^2$ block through riparian strips. Roads should not bisect reserve areas and should cross riparian strips as little as possible. Where roads do bisect reserved areas, bridging points where crowns of trees interlock above the road should be maintained if at all possible.

### Harvesting inventory

This normally takes place at least 6 months before the onset of harvesting concurrent with climber cutting (see below). A 100% timber inventory is required in all workable areas of the annual coupe. All commercial trees over the minimum diameter are enumerated and mapped. To facilitate damage limitation during harvesting, felling directions should also be marked at this time to align with the placement of skidroads.

In addition to marking trees for harvesting, advanced growth of commercial trees and certain mature timber and non-timber trees of particular biological value may be marked for retention. However, the practicality of marking for retention will, to a large extent depend on the felling intensity. In conditions of light logging a higher percentage of advanced growth would be expected to survive than under conditions of intensive felling. This will be reflected in envisaged management cycles.

Relatively few mature trees would normally be marked for retention per unit area. Occasional mature or over-mature (hollow or otherwise of poor quality) timber trees might be retained to act as seed trees, particularly if that species's seeds are not long retained in the soil seed bank (as is the case with most climax species) and its seeds are not widely dispersed. It would be particularly important to retain old specimens of rare and valuable trees for this purpose (e.g. *Intsia palembanica* and *Dyera costulata* in peninsular Malaysian hill forests).

Deciding which non-commercial trees to mark for retention for their biological value requires extensive knowledge of the forest ecosystem concerned as well as site-specific data. In some cases, attempts may be made to identify 'keystone' trees. It has been recommended that field operations should attempt to retain large individuals of trees producing sugary fruits of small seed size, such as figs, and used by a large cohort of frugivores.

Reduction in fig density of up to 74% in logged[1-18:51] forests of peninsular Malaysian does not result in reduced densities of many of the frugivores that exploit them (Grieser Johns in press a). Figs may actually be less crucial to the persistence of most frugivores than is often thought. It is perhaps more useful to identify trees that are used by frugivores when figs and similar fruits are not available. For example, in a study in East Kalimantan, Indonesia, Leighton & Leighton (1983) demonstrated that large-seeded capsular fruits of the Meliaceae and Myristicaceae, and woody climbers of the Annonaceae, may be particularly important in providing resources for frugivores during the few periods when figs are not available. In the peninsular Malaysian study, trees of the Meliaceae and Myristicaceae used by large frugivores were reduced in density by an average of 60%. In this case, this did not affect the density of frugivores any more than a loss of figs, but at high damage levels it is within these taxa that co-relationships are more likely to occur. Persistence rates of potential 'keystone' trees could be improved through marking for retention and the damage limitation techniques described below.

Deciding the density of snags to mark for retention is also difficult. In temperate forests of North America, retention of 5–10 snags/ha is usually recommended, although in most cases the density of snags or of refuge/breeding cavities has not been correlated with the density of animal species dependent upon them (e.g. Welsh & Copen 1992). The resource appears over-abundant in most forests. On the other hand, total elimination of the resource is to be avoided, as is demonstrable under-provision of the resource (retention of only 1.5 snags/ha in some southern Australian clear-felled forests is clearly inadequate: Lindenmayer *et al.* 1990a).

Snags typically occur at a density of around 10–20/ha in tropical forest and cavities at densities of 60–70/ha. The loss of 50% of snags does not appear to cause drops in the density of birds using them as foraging and/or breeding sites in peninsular Malaysia or Sabah (Grieser Johns 1996, in press b). A loss of 50% of cavities should not affect bird numbers in Ugandan forest since ≤ 50% are occupied (Dranzoa 1995). It may thus be sufficient to mark 50% of large snags for active retention during the harvesting inventory, or ensuring a retained density of ≥ 5/ha, whichever is greater.

### Climber cutting

Interlinking of trees by climbers can result in many additional trees being pulled down by felled trees during harvesting. Furthermore, some 20% of climbers are strong enough to remain unbroken when a tree is felled, which can mean that other trees have to be cut to get the commercial stem to the ground. Two studies in Malaysian dipterocarp forest have shown that advance cutting of climbers reduces the number of trees pulled down by 16% and 45% (Fox 1968; Appenah & P··tz 1984). In Suriname, cutting climbers 6 months in advance of felling resulted in only 16% of felled trees pulling over others (Hendrison 1990).

Cut climbers will coppice unless poisoned, and cut ends will often re-root themselves. Most of the climbers that are not cut will survive the felling of host trees. Typically, fewer trees in logged forest support climbers (e.g. a reduction from 26% in unlogged to 12% of trees in logged[28–36:?] forest at Sungai Lalang, peninsular Malaysia: Laidlaw 1994), but this is primarily a consequence of the large numbers of fast-growing pioneer trees in the logged forest sample. The species diversity of climbers is not much affected by cutting prior to tree felling, and it is generally to be recommended. However, in conditions of excessive canopy opening the growth of leafy climbers in recently logged forest can be so vigorous as to retard the regeneration of commercial trees and may become a management problem (e.g. *Mezoneuron* in Sabah and *Merremia* in the Solomon Islands: Neil 1984; Whitmore 1984). Biodiversity is also likely to be reduced where climbers repress normal regeneration processes. Management needs to be directed towards the prevention of climber infestation in heavily damaged logged areas.

### Harvesting procedures
#### Roading
Planning of the road network should take into account potentially unstable areas (swampy depressions, steep slopes prone to landslips, etc.). Drainage lines (including channels which carry water only during storms) need to be avoided or properly bridged. It may be appropriate to divide the road network into all-weather and dry-weather sections to achieve a compromise between costs of building all-weather roads and unacceptable levels of erosion. All-weather roads will need clearance of 20-m wide strips on each side to facilitate drying out after rain, or will need careful preparation of the road surface.

Construction of principal roads should be timed to avoid periods of heavy rainfall and to allow sufficient time for the roadbed to stabilize before it is subjected to heavy traffic. Adequate drainage needs to be provided. The use of excavators rather than bulldozers has been recommended for building forest roads because of their more accurate placement of soil and a subsequent lesser level of incidental damage to trees (Marn 1982).

Location of skidroads should also be planned, based on the densities and positions of commercial stems. Where possible, skidroads should be located between drainage lines with the aim of skidding logs uphill rather than downhill (in steep country this is generally practical only in cable logging operations). This would mean that water flows into progressively less disturbed areas as it moves downhill and much of the carried sediment may be filtered out before it reaches valley bottom streams. Skidroad networks orienting downhill are more prone to concentrating downslope waterflow into roadlines, causing considerable gulley erosion and a high level of silt dumping into the streams.

Apart from in Australian forests, where close supervision has been observed since the early 1980s, the idealized skidroad pattern projected along water divides is rarely followed in practice (Fig. 9.2). It is important that field operations pay more attention to restricting machines to planned trails and ensuring felled logs are winched to the machine rather than allowing the machine to move to the log. This ensures that the various trees marked for retention are not casually destroyed by contractors.

#### Felling
Felling and the subsequent skidding of logs are the critical phases of logging operations as it is here that most costs, both economic and

environmental, are incurred. In uncontrolled felling operations, trees are cut as they are located without specified procedures or cutting directions. This tends to result in greater damage levels than necessary and the creation of larger gaps. Uncontrolled felling in Suriname led to the creation of gaps 2.5 times the size of the maximum gap in controlled felling areas (Hendrison 1990). Controlled directional felling also maximizes logging efficiency, minimizing both the need for construction of extra skidroad length and the lengths over which logs need to be winched to the skidroads (Fig. 9.3).

Fig. 9.2. Planned skidroad design and actual skidroad construction in a lowland rain forest, Suriname. Planned skidroad design (a) is according to the forest management plan, actual skidroad layout (b) is that established by the contractors. Heavy lines are main access roads. The Mapane River drainage system is illustrated. (Source: Hendrison 1989.)

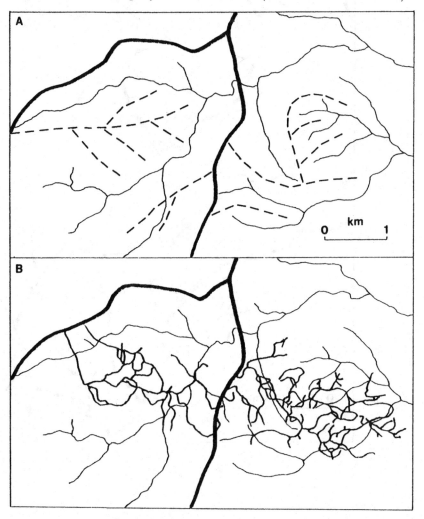

Fig. 9.3. Uncontrolled (a) and controlled directional felling patterns (b). (a) Bia Forest, Ghana; (b) Mapane, Suriname. In the Bia Forest, felling intensity was 1.2 trees/ha and damage level around 27%. At Mapane, felling intensity was 3.8 trees/ha and damage level 17.3%. Average log size was larger at the Bia Forest. Variable width lines are skidroads. The approximate (Bia Forest) or known (Mapane) position of felled trees is illustrated. (Sources: Bia Forest adapted from Hawthorne 1993; Mapane after Hendrison 1989.)

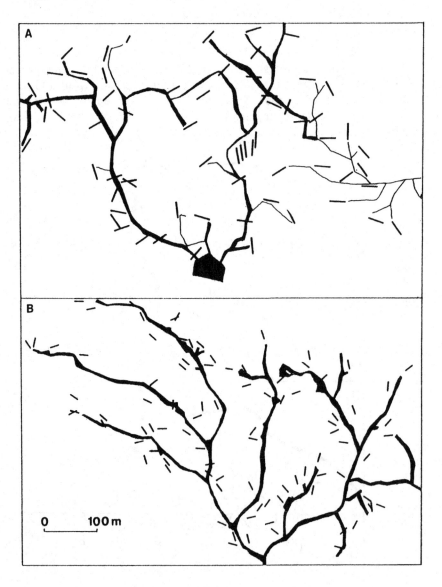

Directional felling requires training of chainsaw teams, provision of felling wedges, and close observance of safety rules. The direction in which rain forest trees can be felled depends to some extent on the shape and size of the tree crown and the occurrence of lianas connecting the crown to adjacent trees. The latter is reduced by prior cutting of climbers, but judgement still needs to be exercised to assess the natural lean of the tree (it is possible to fell within about 45° of the centre of the direction of lean) and the most appropriate felling direction in relation to the skidroad pattern. Ideally, felling should aim for a herringbone pattern: the felling direction should be 30–60° to the nearest skid trails, bearing in mind the direction in which logs will be extracted (Hendrison 1990). In felling, attention should also be paid to where the crown will land and ideally the tree should be felled towards relatively open areas or forest poorly stocked with commercial or other trees marked for retention. Trees which cannot be felled in an appropriate pattern, either because of lean or because of the danger of damaging important marked trees, can be cut into shorter lengths to make turning the logs less damaging to the surrounding vegetation. Trees which fail to fall to the ground due to hang-ups should be winched down by skidders rather than lowered by the progressive felling of other trees.

Felling of low-yield trees with structural defects or heart rot should be avoided. Close inspection to determine structural defects, or rotten and hollow trunks, can avoid unnecessary felling and the tree can be left as a seed bearer. Isolated commercial trees standing some distance from the skidroad network should also be left since construction of a spur to extract a single tree is generally uneconomic, particularly if damage to residuals is taken into account.

### Extraction

In many conventional logging operations there is no planned layout of skidtrails prior to extraction of logs. Skidding machines move into the forest until they locate a log, then drag it to a loading point by the shortest route. This results in damage levels 2–3 times as high as where skidding is controlled. It is preferable to lay out the skidroads before felling, clearing primary trails and marking any necessary branches such that skidder drivers have clear instructions as to where they are permitted to drive. The length of necessary skidtrail branches can be reduced by winching logs from the stump to existing trails, for which logs may have to be cut to manageable lengths.

Damage to the soil along skidtrails can be minimized by ensuring high payloads for each trip by the skidder. Frequent trips cause more damage than fewer trips with heavier loads. Chokers can be used to attach more than one log to the machine up to its hauling capacity (Marn 1982; Malvas 1987). Where the design of the machine allows (i.e. where an integral logging arch is installed on the winch) the butts of the logs should be kept off the ground to reduce drag resistance and avoid trenching which on slopes is especially prone to develop erosion channels. Skidding during wet weather should be avoided wherever possible.

Marked residual trees along skidding trails should be protected by laying a log guide or protection log on the exposed side. This prevents buttress or bark damage by hauled logs which can lead to fungal infection of the damaged tree.

### Post-felling procedures
#### Clean-up procedures
Removal of badly damaged trees of commercial species with a marketable log is allowable, even if below minimum felling diameter. In some areas, such as southern Uganda, discarded stumps and brash have in past operations been converted into charcoal in a follow-up to the commercial felling operation (Earl 1968). This has the effect of removing a large proportion of wood biomass from the forest, with some associated loss of nutrients. There is no evidence as to whether this loss significantly affects forest productivity (Dranzoa & Johns 1992), but it is inevitably associated with high levels of environmental damage requiring extensive intervention.

As concession personnel leave the coupe, all discarded cables, oil drums, plastic shelters and other rubbish needs to be collected and removed. Water collecting in drums and other containers can trap and kill small mammals and other animals, as well as providing a breeding site for various insects not normally occurring in primary tropical forest.

#### Water run-off control
Compacted soils on log landing areas and major skidroads should be ripped (using a skidder with attachable ripper teeth fitted to the pushblade) and re-seeded (restored) where necessary. Roadside earthworks and quarries where road surfacing material has been removed should be re-seeded. Re-seeding should use pioneer species of native tree species where possible: the use of exotics has frequently been unsuccessful in the long term despite promising initial trials (e.g. *Acacia mangium* in peninsular Malaysia). Where skidroads have been built up slopes, cross-drains should be constructed to allow run-off without erosion of the roadbed.

Temporary stream crossings should be removed to prevent clogging, and if dams and water back-up have occurred at any point the blockage should be removed.

### Post-felling stand inventory

Some time after the harvesting event the remnant stand should be inventoried to determine the need for interventions, such as enrichment planting of timber or other trees, or further restoration work. Post-felling treatments which involve the refining of the stand are justifiable in some cases to improve the timber stocking (e.g. in the CELOS system of Suriname: de Graaf 1986), providing this does not reduce the overall diversity of forest vegetation. The use of chemical arboricides should be minimized and care taken in their handling and application.

Such work is economically very expensive. Care taken during the harvesting phase should negate the need for excessive interventions if sufficient advanced growth of economically viable timber species is present. A high degree of alteration of the species composition of the forest is akin to the establishment of hardwood plantations. This would be open to all the problems associated with plantations of natural hardwoods. Should such plantations actually be viable they may be better located on land already deforested (although land-tenure problems may argue against this).

### Post-felling biodiversity inventory

This is not a standard part of forest management procedures at the present time but has been recommended in guidelines produced for ITTO (Blockhus *et al.* 1992). The function, as defined in these guidelines, is to determine the need for remedial action when the loss of biodiversity in managed forest is unacceptably high. However, the level of loss that is 'unacceptable' has to be defined locally, as does the form that remedial action might take.

The periodic post-felling biodiversity inventory conducted at the Tekam Forest Reserve, peninsular Malaysia, is an example of the format that such work could take (Johns 1989a). This is conducted in parallel to timber inventory work and is directed towards the rapid assessment of mammal and bird species present and relative numbers. It is conducted every 6 years and takes approximately 2 weeks (100 man-hours) to survey each coupe (four sample coupes in this case). Inventory programmes of this type rely on detailed baseline information collected in unlogged forest and on a fairly high level of expertise in field personnel, but are capable of generating the type of data from which assessments of the biological value of the regenerating forest can be determined objectively. It is important to note that only long-term monitoring (i.e. not inter-site comparisons) can provide usable data concerning requirements for specific interventions (see Chapter 7).

If circumstances allow, collection of more detailed information might be considered, particularly of invertebrate taxa and abundance of certain invertebrate groups, such as dragonflies or papilionid butterflies in any tropical forest (Sutton & Collins 1991) or arctiid moths in dipterocarp forest (Holloway 1984b). However, investigators need to be aware of the ecological role of the targetted taxa (low numbers rather than high numbers can indicate progress in forest regeneration) and the general pitfalls involved in the use of indicator groups.

### Rules and regulations

Most countries have fairly detailed forestry rules. Perhaps the most comprehensive, and arguably the best, were developed for the Queensland forests and imposed with effective penalty structures from the early 1980s, resulting in a dramtic improvement in the standard of forestry practice (Cassell *et al.* 1984). The main points of this legislation were as follows:

The concessionaire and the forest authorities must first agree on definitions of soil type and stability in the concession area. A joint silvicultural enumeration and topographic survey is then carried out to develop a mutually acceptable logging plan. This plan contains maps where all exclusion zones (including riparian strips and catchment areas) and other special management zones are indicated. Main access roads and skidtrails are located according to the environmental criteria discussed above, and maximum trail widths and maximum allowable gradients are specified. Types of drainage work needed are specified: for example, the distance between cross drains along roads and major skidtrails is put at 10–60 m depending on soil type. Instructions exist as to the minimum standards for stream and river crossings, including temporary crossings.

Logging operations are allowed only outside of the main wet season, and all drainage work into new coupes must be completed before the onset of these annual rains. Felling and skidding is closely supervised and only certain types of logging equipment is allowed. The construction of secondary skidtrails to logs is allowed only if winching the log to main skidroads is not possible: all skidders must be equipped with winches and arches capable of lifting the front end of the log during skidding. During logging, spillage of diesel or fuel oils must be avoided to avoid pollution. After logging is complete all non-organic debris must be removed from the coupe.

The forest authority has the right to suspend the logging operation at any time if work is substandard or if it affects water quality. If specified operations to minimize the environmental impact of the operation are not undertaken, the authority may employ another contractor to do the job and charge it to the concessionaire. Fines are charged for trees damaged by the felling and skidding operation, with fines varying according to the size and species of tree.

In return for adherence to these rules on the part of the concessionaire, the forest authority undertakes to guarantee a market for extracted logs and to provide staff trained to supervise all aspects of the logging operation. The latter is a crucial point.

Outside Australia, two main problems are apparent. First, a failure of most concessionaires to observe rules laid down by forest authorities (either wilfully or through a lack of suitably trained field personnel). Second, an inability of the authority to intervene to suspend poorly conducted operations. Together, these have resulted in the failure of management operations that have proven promising in closely supervised pilot studies.

There are many examples of forestry experiments, demonstrations or research plots, where close supervision of logging operations has indicated how the environmental impact of logging operations can be reduced and possibilities for sustainable timber production enhanced. These often provide economic justification for the adoption of best practice. These experiments range from village-scale forestry operations (such as the Palcazu demonstration in Peru: Hartshorn 1990) to experimental commercial operations (such as the CELOS project in Suriname: de Graaf 1986). Other examples concern dipterocarp forest in Sarawak (Marn 1982), the Tapajós Forest Reserve in Brazil (UNDP/FAO 1983) and several forest types in Côte d'Ivoire (Maître 1987). The point has been made time and time again that reduced-impact logging is viable and desirable. What is required at the present time is not further reduced-impact experiments but strengthening of the supervisory ability of forest departments and wider application of the rules that already exist.

Compliance with rules, and with modifications or interventions suggested in situ, is essential if tropical forest logging operations are to be improved. It has so far proved difficult to overcome resistance to change among concessionaires, even where suggested changes have minimal or even negative associated costs. Changes need to be enforced rather than voluntary, at least initially. In the case of Queensland, the imposition of an effective legislation was accompanied by an educational programme addressing both the rationale of the legislation and the practical problems of implementation.

Market forces might also play a useful role in stratifying the prices paid for timber produced under various standards of management, or by insisting on certain minimum standards in any timber traded internationally.

### Examples of current 'best practice' management systems

Measures of 'best practice' for forestry operations have been drawn up by ITTO (1990). Adherence to these is now required by most major lending institutions, such as the World Bank, if support for the forestry sector is requested. Best practice involves observation of various methods designed to reduce the environmental impact of logging operations (see discussion above). The intention is to encourage the adoption of forest policy directed towards the sustainable development of the resource (although precisely what will be sustainable remains a subject for conjecture).

Sustainability of current forest harvesting systems is hard to define or demonstrate, primarily because most systems now widely used have evolved over a relatively short space of time and because a sufficient number of rotations have not been carried through to determine their success. Some long-term monitoring of tree regrowth and increment is quite well advanced, however (e.g. at the Tapajós Forest Reserve, Brazil: Silva *et al.* 1995).

The theory behind some polycyclic systems is good, and forestry procedures and rules well thought out (if not necessarily adhered to in the field), and these may prove successful in the long term. Preliminary indications based on the measurements of upper canopy trees suggest that the polycyclic Selection Management System practised in Malaysia since about 1970 has the potential to provide a sustainable supply of timber (World Bank 1991), although no conclusions can yet be drawn about the sustainability of the forest ecosystem as a whole. The monitoring programme established at Tekam Forest Reserve may provide these answers in time.

Where offtakes and associated damage levels are low, such as in remote areas of Zaire where forest management plans call for an offtake of 0.5 trees/ha on a 40-year cycle, selective management systems are unlikely to affect the ecosystem in the long term. That is, they are likely to be sustainable in the widest sense, although no proof of this exists. Many selective management systems involve some level of intervention to increase the representation of commercial species (often arboricidal treatment of non-commercials). This will clearly affect the ecosystem to some extent, narrowing the sense in which operations may be sustainable.

Perhaps the most interesting early example of a potentially sustainable polycyclic system is the periodic block system employed in Trinidad since 1948. This evolved primarily to supply a local timber market and involves the cutting of around $25\,m^3$/ha on a 30-year cycle. Trees to be cut or retained are individually selected on the basis of a number of factors, including frequency in the forest block, usefulness for steam bank protection and for wildlife, and value as a seed tree. The stand characteristics of the forest managed under the periodic block system closely mimic those of natural forest, suggesting that the age structure of the tree population is not significantly affected. Forests which have been managed over more than one cycle are reported to have maintained their biodiversity (Clubbe & Jhilmit 1992), although it is not clear how this has been monitored. Success in maintaining biodiversity may be at least partly due to the fact that the forests are floristically simple, with perhaps no more than 20 tree species growing to a diameter $\geq 19.5\,$cm, but the management principles are sound.

In Myanmar, teak forests have been harvested since the mid-nineteenth century using elephants and water transport. Since elephants can transport only small logs, only intermediate size-classes have been felled. Damage levels are fairly low and large seed trees remain. This so-called Brandis system is prone to degradation if the large canopy trees are not replaced, but if a certain proportion of commercial stems are left to grow into any canopy gaps the sustainable production of timber can continue. Managed forests of this type typically contain even-aged cohorts of regenerating trees under an established canopy. They may lack the vegetational structural diversity typical of mature forest, but they probably contain much of the original biodiversity. No recent studies have taken place in Myanmar, however, so there is no confirmation of this.

### Summary

Reduced-impact logging is essential in the improvement of tropical forest management for both timber production and biodiversity conservation. Means of improving field operations have been well illustrated but have rarely been applied on a large scale. This is due mainly to a lack of trained field staff, an inability of forest authorities to enforce the rules, and a lack of incentives for logging contractors to apply the rules.

Before improved field practices can be applied, the conditions need to be established whereby rules can be observed. If this is the case and if forest policy is to include additional elements of biodiversity conservation, a variety of interventions can be applied.

Small-scale experimental logging operations have indicated corrections to field procedures that should be applied, and there is a general consensus on these. On a wider scale, improved field operations have been successfully tested in Queensland forests. Although none of the experimental operations has included an element of biodiversity monitoring, the application of rules reducing the impact of logging on the environment and thus increasing the potantial for regeneration of the timber crop will equally well benefit biodiversity.

Examination of the persistence of biodiversity alongside some established 'best practice' logging systems, such as those of Trinidad or Myanmar, might provide some evidence as to whether reduced-impact logging may be sustainable in the wider biological sense as well as in terms of timber production.

# 10

## The future

### Integrated management of tropical forests

*Management priorities*

Land use plans for tropical countries need to take into account the needs of (often expanding) rural populations. This is generally paramount in development programmes. Where the proportion of forest land is low compared with population density, forests will need to be shown to be economically productive, either through direct utilization or in providing jobs. Protected areas can only be justified to local populations if they generate revenue from ecotourism or other sources, or if they provide clearly demonstrable benefits as water catchment areas, etc.

Most forest areas in the tropics will be utilized in some way and the challenge for development programmes is to ensure that forms of utilization are sustainable. This has rarely been the case to date (Kremen *et al.* 1994). In many regions community use of forest products is longstanding, and future programmes will need to concentrate on its development and regulation. In other areas, the development of agroforestry or community forestry programmes may need to be encouraged in order to provide sources of fuelwood and construction timber that may not be available from reserved forests. The control of commercial forestry is rarely vested in local communities and this may be a requirement for change.

In less populated countries, it may be possible to preserve large areas of tropical forest providing some benefits accrue to the administrative authorities. A percentage of required hardwood timber may be provided from plantations established on degraded or non-forest land or compensatory payments may be made for retaining forest for its value as a carbon sink (Sedjo 1989; Panayotou & Ashton 1992) or for less easily calculated global benefits such as the preservation of biodiversity.

The setting aside and active policing of large areas of intact forest would ensure the preservation of tropical biodiversity. In most cases, however, this is not a viable land-use option. A more appropriate form of management is the preservation of small undisturbed forest areas within a larger matrix of production forest. Ideally, production zones should be zoned with respect to their distance from preserved areas and proximity to local communities (who may require areas in which hunting, collection of fuelwood, and other such activities are permitted). Fuelwood production zones may need to be created on the edge of natural forest reserves, using fast-growing and coppicing species, as has been suggested for the Mount Elgon National Park and various other reserved forest areas in Uganda (D. Earl, pers. comm.). The innermost production forest areas, surrounding totally protected areas, may be managed for production of timber and other forest products, but management should follow clear guidelines laid down to enhance the conservation value of the regenerating forest.

### Management for biodiversity

Until the beginning of the 1980s it was a maxim of practical conservation that areas of disturbed habitat were not considered useful in maintaining tropical biodiversity and were incorporated into conservation strategies only where undisturbed habitat was critically reduced or non-existent (Johns 1983). This is exemplified by the fact that tropical forest degraded by logging was generally lumped together with totally deforested areas in statistical analysis of land-use trends (e.g. Myers 1989). Johns (1983) first illustrated the high biodiversity retained in a tropical rain forest managed primarily for timber production and the potential for integrating conservation criteria within evolving improved forest management systems.

By the mid-1980s, international conservation agencies were beginning to develop ecological guidelines for the management of tropical forest lands within which the potential of logged-over forests for biodiversity conservation was considered (Poore & Sayer 1987). However, forest policies determined by foresters themselves still tend to acknowledge the importance of retaining biodiversity without actually considering interventions to increase habitat value (FAO 1993). In the priorities of lending institutions, management to benefit biodiversity is generally mentioned only as a potential byproduct of other policies (e.g. UK Government 1994).

There is a frequently stated need for research to determine ways of conserving biodiversity as an integral part of tropical forest management, but little evidence that this research is actually being carried out. Some broad interventions have been suggested in guidelines produced for ITTO (Blockhus *et al.* 1992) but these either follow general principles of low-damage polycyclic management systems (particularly the Queensland selective management system) or are too general to be applied directly to forest management planning. The guidelines are useful only as a framework for local interpretation by suitably trained foresters. In general, there is a lack both of manuals of appropriate field practice for forests of different types and of foresters trained to interpret them (Palmer & Synnott 1992). No existing manual for field practice contains elements directed specifically towards the conservation of biodiversity.

Priorities for the incorporation of biodiversity conservation into forest management strategies may change as the timber industry spreads into remaining pristine areas. Johns (1989a) pointed out that research and development of sustainable management strategies will sooner or later run out of undisturbed forest estate and that experimentation will then begin to be conducted within National Parks or other preserved areas. Gómez-Pompa and Burley (1991) subsequently suggested that certain natural regeneration logging systems would in any event be appropriate for National Parks and other conservation areas where exploitation could benefit local inhabitants, biodiversity could be maintained and the protected status of the forest would ensure integrity of the forest management unit. This appears to ignore certain central tenets of National Park (protected areas) management policy and would be considered undesirable by many foresters and almost all conservationists. However, it needs to be recognized that there is a growing trend for the incorporation of (generally undefined levels of) sustainable use by local populations into protected area management in the tropics, up to and including extraction of timber (e.g. Howard *et al.* 1995). Although local rather than commercial use of timber is generally intended, these can be equally intensive.

In the short term, given forest management systems employed and possible future modifications, it is possible to predict broad changes in plant and animal species abundances. Models such as DYNAST and FORPLAN have been developed for temperate forests which compare multiple-use benefits, including maintenance of animal species populations, under different management strategies (Benson & Laudenslayer 1986; Burgman *et al.* 1994; McCarthy & Burgman 1995). While no similar models have been developed for or applied to tropical forests, the implications for biodiversity conservation of various alternate scenarios of preserved and production forest areas can be estimated using results from field studies, particularly those involving long-term monitoring. Species likely to be immediately threatened by specific forestry practices can be identified.

In the long term it is difficult to be specific, but it is clear that the loss of large preserved forest areas may result in erosion of genetic diversity of some species if not necessarily their local extinction. On the other hand, improved management of production forest to protect residual trees and enhance natural regeneration is likely to increase the value of production areas for biodiversity. The most important factor affecting long-term trends will be the efficiency with which forestry rules are observed and the ability of foresters to protect the integrity of the forest estate (Crome *et al.* 1992).

### Species extinctions

Models exploring and predicting the loss of tropical species have concentrated on deforestation as the critical factor and have estimated species loss rates as high as 14% by the year 2015 and 40% by 2040 (Reid 1992). However, evidence from field studies suggests extinctions have so far been minimal, despite extensive deforestation in many biogeographical regions (Heywood & Stuart 1992).

Models linking deforestation with species extinctions have proved inaccurate due to the definition of deforested land employed. Forests that are not clear-felled but are degraded in some way, usually by shifting cultivation or timber logging, are considered 'deforested'. Such areas often occupy considerable portions of biogeographic regions and continue to support at least part of the original biodiversity. A combination of the existing preserved areas and degraded habitats have so far succeeded in preserving most species, although many are considered to exist at critically low population levels and thus to be in danger of extinction. Loss or alteration of habitat through human settlement and agriculture is commonly given as a cause for critical declines in species populations (e.g. Collar *et al.* 1994). However, no species of plant or animal has yet been reported as becoming extinct as a result of timber production from tropical forests (Table 10.1).

Among mammals and birds, forestry activities have been reported as the sole threat or cause of decline in four bird species and a principle cause of decline in an additional 10 birds and five mammals (Table 10.1). The bird species reported as threatened solely by logging are the imitator sparrowhawk *Accipiter imitator*, yellow-legged pigeon *Columba pallidiceps*, and chestnut-bellied kingfisher *Todirhamphus farquhari*, all of southern Melanesia, and the white-tipped monarch *Monarcha everetti* of Sulawesi (Collar *et al.* 1994). It is worth noting that these are all island forms and that the effects of forest management practices within the less diverse island ecosystems have not yet been studied. The other affected bird species cover a range of taxa and geographical locations, although two more are large-bodied frugivorous pigeons. The five affected mammals are all African primates and three are lemurs.

Table 10.1. *Causes of species extinctions since 1600 and reasons for threatened status of tropical forest species considered Critical, Endangered or Vulnerable (IUCN categories)*

| Cause of extinction or threats | Extinctions[c] | | Threats (tropical forest species) | |
|---|---|---|---|---|
| | Mammals | Birds | Mammals | Birds |
| **Island endemics[a]** | | | | |
| Introduced diseases 2 | 0 | 0 | 3 | |
| Introduced diseases plus habitat loss | 0 | 12 | 0 | 8 |
| Introduced predators | 2 | 41 | 1 | 27 |
| Introduced predators plus habitat loss | 13 | 13 | 2 | 9 |
| Introduced predators plus hunting | 0 | 7 | 0 | 0 |
| Hunting | 0 | 5 | 3 | 12 |
| Hunting plus habitat loss | 0 | 0 | 4 | 53 |
| Hunting plus habitat degradation by logging | 0 | 0 | 0 | 1 |
| Habitat loss | 0 | 1 | 1 | 88 |
| Habitat loss plus degradation by logging | 0 | 0 | 0 | 3 |
| Habitat degradation by fire | 0 | 0 | 0 | 1 |
| Habitat degradation by logging | 0 | 0 | 0 | 4 |
| Natural causes or unknown | 0 | 0 | 0 | 19 |
| **Continental species[b]** | | | | |
| Introduced predators | 1 | 0 | 1 | 0 |
| Introduced predators plus habitat loss | 3 | 0 | 0 | 0 |
| Introduced predators plus habitat degradation by grazing of rangelands | 5 | 0 | 0 | 0 |
| Hunting | 11 | 0 | 10 | 5 |
| Hunting plus habitat loss | 1 | 4 | 44 | 37 |
| Hunting plus habitat degradation by logging | 0 | 0 | 1 | 1 |
| Habitat loss | 0 | 8 | 27 | 200 |
| Habitat loss plus degradation by logging | 0 | 0 | 4 | 5 |
| Habitat degradation by logging | 0 | 0 | 0 | 0 |
| Habitat degradation by fire | 0 | 0 | 0 | 2 |
| Habitat degradation by grazing of rangelands | 1 | 0 | 0 | 0 |
| Habitat degradation by draining of wetlands | 0 | 3 | 0 | 0 |

Table 10.1. (*cont*).

| Cause of extinction or threats | Extinctions[c] | | Threats (tropical forest species) | |
|---|---|---|---|---|
| | Mammals | Birds | Mammals | Birds |
| Food resource competition (with humans) | 0 | 1 | 0 | 0 |
| Natural causes or unknown | 3 | 2 | 1 | 8 |
| Total | 42 | 97 | 99 | 486 |

Figures are approximations only as the status of species is constantly changing.
[a]Includes Philippines and New Zealand.
[b]Includes Australia, Borneo, Sumatra, Java and New Guinea.
[c]Includes species whose wild populations are extinct but which are maintained in captivity. Excludes island species known from bones and other remains (e.g. 25 mammal species reported from the Caribbean that may have persisted beyond 1600 and as many as 200 species of birds from the Pacific).
*Sources*: IUCN data sheets or summaries (numerous publications).

Total numbers of tropical forest species objectively reported as directly threatened by logging is currently minimal. This is to some extent a reflection of conservation criteria: a species vulnerable to logging in the ecological sense is only classified as 'Vulnerable' in the conservation sense if little unlogged habitat remains within its geographical range. However, many species which might otherwise be listed as of conservation concern are not because of their ability to persist in logged or otherwise degraded forests. The conservation value of properly managed timber production forests can be high, and can be improved still further. However, the point must be made that poorly managed timber forests create a threat to many more species than are mentioned above through encouragement of deforestation. The extent to which long-term management is applied is a more important criterion in many cases than the form of logging.

### A conservation ethos

By its nature, forestry is concerned with maintaining the quality of the environment and with the conservation of various non-marketed benefits (Earl 1973). Forestry is certainly not an isolated science: Wilson (1993) considers that forestry is merely a branch of ecology and that the development of tropical forest management links closely with the development of tropical ecology and associated studies of biodiversity.

Conservation values do not need to be introduced into forest management strategies because they are already there. Foresters, although not necessarily logging contractors, are by definition trained in the conservation and management of a natural resource. Correctly managed production forests conserve hydrological and climatic functions, conserve tree and other plant species-richness on a concession-wide scale (particularly where planning includes some provision of reserves) and allows the survival of many animal species without special precautions. Specific well-researched interventions can conserve still more of the biodiversity without a significant reduction of cost-effectiveness.

Conservation training within forestry curricula can identify the convergence between the long-term needs of forestry and biodiversity conservation, explain the rationale for conservation interventions and provide further emphasis to the vital requirement for observance of existing environmental rules. The important role of foresters in the development of national conservation strategies should be emphasized. While the training programmes that produce environmentalists are more difficult to define, a certain amount of education concerning the potential importance of forestry as a conservation tool might equally well be applied.

As long ago as 1970, it began to be suggested that foresters should reconsider the management of forests as public amenities rather than industrial production areas (Richardson 1970). While this caused a certain amount of dissent at the time, the responsibilities of foresters are certainly changing. The foresters of the future can expect to be as concerned with managing preserved areas (such as VJRs or equivalent nature reserves), and multiple-use areas (where ecotourism, hunting or community use of non-timber products is practised) as they will be with managing timber production. Strict nature reserves, which have more rigorous protection criteria than any other preserved areas, including National Parks, are frequently located within the forest estate and managed by Forest Departments (e.g. in Madagascar: IUCN 1986). There is a clear precedent for the integration of forest preservation and management of the economic resource: these are all part of a sustainable forest management strategy.

### Summary

Most tropical countries posses rapidly expanding rural populations. Retained areas of tropical forest are under increasing pressure to prove themselves an economically viable land-use option. Production forestry is generally demonstrably more economic than forest preservation for its amenity values.

In recognition of this fact, there is an increasing trend towards the incorporation of sustainable use, including timber and fuelwood collection, into forest National Parks and other preserved areas. In the future it is likely that preserved forest will exist only as fragments within a larger matrix of production forest. It is currently of extreme importance to research forms of use that are genuinely sustainable. It is also important to zone production areas according to their proximity to human settlements and likely intensity of use, and to research and apply means of enhancing the value of at least the innermost zones for biodiversity conservation.

The potential value of production forest has been underestimated. Until comparatively recently it has been considered of little value for biodiversity conservation. In fact, no species has yet become extinct as a direct result of tropical forestry operations. As pressures on preserved areas grow, foresters may hold the key to the conservation of tropical biodiversity.

# REFERENCES

Abbott, I. & Heurck, P. U. (1985). Response of bird populations in jarrah and yarri forest in Western Australia following removal of half the canopy of jarrah forest. *Australian Forester*, **48**, 227–34.

Abdulhadi, R., Kartawinata, K. & Sugardjo, S. (1981). Effects of mechanized logging in the lowland dipterocarp forest at Lempake, East Kalimantan. *Malaysian Forester*, **44**, 407–18.

Alaric Sample, V., Johnson, N., Aplet, G. H. & Olson, J. T. (1993). Introduction: defining sustainable forestry. In *Defining sustainable forestry*, ed. G. H. Aplet, N. Johnson, J. T. Olson & V. Alaric Sample, pp. 3–8. Washington, DC: Island Press.

Alexander, I., Norani Ahmed & Lee, S. S. (1992). The role of mycorrhizas in the regeneration of some Malaysian forest trees. *Philosophical Transactions of the Royal Society of London B*, **335**, 379–88.

Ambrose, G. L. (1982). An ecological and behavioural study of vertebrates using hollows in eucalyptus branches. PhD thesis, La Trobe University, Melbourne.

Amir, H. M. S., Zakaria, M., Ghazali Hasan, M. & Ahmad, R. (1990). Nutrient dynamics of Tekam Forest Reserve, peninsular Malaysia, under different logging phases. *Journal of Tropical Ecology*, **2**, 71–80.

Amo, R. S. del (1991). Management of secondary vegetation for artificial creation of useful rain forest in Uxpanapa, Vera Cruz, Mexico: an intermediate alternative between transformation and modification. In *Rain forest regeneration and management*, ed. A. Gómez-Pompa, T. C. Whitmore & M. Hadley, pp. 343–50. Paris: UNESCO.

Appenah, S. & Putz, F. E. (1984). Climber abundance in virgin dipterocarp forest and the effect of pre-felling climber cutting on logging damage. *Malaysian Forester*, **47**, 335–42.

Arnold, G. W., Steven, D. E., Weeldenburg, J. R. & Smith, E. A. (1993). Influences of remnant size, spacing pattern and connectivity on population boundaries and demography in euros *Macropus robustus* living in a fragmented landscape. *Biological Conservation*, **64**, 219–30.

Asabere, P. K. (1987). Attempts at sustained yield management in the tropical high forests of Ghana. In *Natural management of tropical moist forests*, ed F. Mergen & J. Vincent, pp. 47–70. New Haven: Yale University Press.

Ashton, P. S. (1976). Factors affecting the development and conservation of tree genetic resources in South-east Asia. In *Tropical trees: variation, breeding and conservation*, ed. J. Burley & B. T. Sykes, pp. 189–98. London: Academic Press.

Asibey, E. O. A. (1978). Primate conservation in Ghana. In *Recent advances in primatology*, vol. 2, ed. D. J. Chivers & W. Lane-Petter, pp. 55–9. London: Academic Press.

Attiwill, P. M. (1994a). The disturbance of forest ecosystems: the ecological basis for conservation management. *Forest Ecology and Management*, **63**, 247–300.

Attiwill, P. M. (1994b). Ecological disturbance and the conservation management of eucalypt forests in Australia. *Forest Ecology and Management*, **63**, 301–46.

Ayres, J. M. (1981). *Observacoes sobre a ecologia e o comportamento dos cuxiús* (*Chiropotes albinasus e Chiropotes satanas, Cebidae: Primates*). Belém: Grafisa.

Ayres, J. M. (1986a). Some aspects of social problems facing conservation in Brazil. *Trends in Ecology and Evolution*, **1**, 48–9.

Ayres, J. M. (1986b). Uakaris and Amazonian flooded forest. PhD thesis, University of Cambridge.

Ayres, J. M. & Johns, A. D. (1987). Conservation of white uacaries in Amazonian várzea. *Oryx*, **21**, 74–80.

Ball, J. B. (1992). Forest plantations and the wise management of tropical forests. In *Wise management of tropical forests*, ed. F. R. Miller and K. L. Adam, pp. 97–109. Oxford: Oxford Forestry Institute.

Barbier, E. B. (1993). Economic aspects of tropical deforestation in South-east Asia. *Global Ecology and Biogeography Letters*, **3**, 1–20.

Barbier, E. B., Burgess, J. C., Aylward, B., Bishop, J. T. & Bann, C. (1993). The economic linkages between the trade in tropical timber and the sustainable management of tropical forests. Final report to project PCM(XI)/4, ITTO, Yokohama.

Barbier, E. B., Burgess, J. C., Bishop, J. & Aylward, B. (1994). *The economics of the tropical timber trade*. London: Earthscan Publications.

Barrett, E. B. M. (1984). The ecology of some nocturnal mammals in the rain forest of peninsular Malaysia. PhD thesis, University of Cambridge.

Barton, A. M. (1984). Neotropical pioneers and shade- tolerant tree species: do they partition treefall gaps? *Tropical Ecology*, **25**, 196–202.

Beaman, R. S., Beaman, J. H., Marsh, C. W. & Woods, P. V. (1985). Drought and forest fires in Sabah in 1983. *Sabah Society Journal*, **8**, 10–30.

Beehler, B. M., Krishna Raju, K. S. R. & Ali, S. (1987). Avian use of man-disturbed forest habitats in the Eastern Ghats, India. *Ibis*, **129**, 197–211.

Bell, H. L. (1982). A bird community of lowland rainforest in New Guinea. III. Vertical distribution of the avifauna. *Emu*, **82**, 143–62.

Benkman, C. W. (1993). Logging, conifers and the conservation of crossbills. *Conservation Biology*, **7**, 473–79.

Bennun, L., Dranzoa, C. & Pomeroy, D. (In press). The forest birds of Kenya and Uganda. *African Journal of Ecology*.

Benson, G. L. & Lauderslayer, W. F. (1986). DYNAST: simulating wildlife responses to forest management strategies. In *Wildlife 2000*, ed. J. Verner, M. L. Morrison & C. J. Ralph, pp. 351–5. Madison: University of Wisconsin Press.

Blanche, C. A. (1978). An overview of the effects and implications of Philippine selective logging on the forest ecosystem. *BIOTROP Special Publications*, **3**, 97–109.

Blockhus, J. M., Dillenbeck, M., Sayer, J. A. & Wegge, P. (1992). *Conserving biological diversity in managed tropical forests*. Gland, Switzerland: IUCN.

Bohlman, S. A., Matelson, T. J. & Nadkarni, N. M. (1995). Moisture and temperature patterns of canopy humus and forest floor soil of a montane cloud forest, Costa Rica. *Biotropica*, **27**, 13–19.

Bol, M. M. G. R. & Beekman, F. (1989). Economically and environmentally sound harvesting methods. *Communications of the Norwegian Forestry Research Institute*, **41**, 319–30.

Borhan, M., Johari, B. & Quah, E. S. (1987). Studies on logging damage due to different methods and intensities of forest harvesting in a hill dipterocarp forest of peninsular Malaysia. *Malaysian Forester*, **50**, 135–47.

Bourlière, F. (1989). Mammalian species richness in tropical rainforests. In *Vertebrates in complex tropical systems*, ed. M. L. Harmelin-Vivien & F. Bourlière, pp. 153–68. Berlin, Heidelberg, New York: Springer.

Braithwaite, L. W., Turner, J. & Kelly, J. (1984). Studies on the arboreal marsupial fauna of eucalypt forests being harvested for woodpulp at Eden, NSW. III. Relationships between faunal densities, eucalypt occurrence and foliage nutrients, and soil parent materials. *Australian Wildlife Research*, **11**, 41–8.

Branch, L. C. (1983). Seasonal and habitat differences in the abundance of primates in the Amazon (Tapajós) National Park, Brazil. *Primates*, **24**, 424–31.

Brandani, A., Hartshorn, G. & Orians, G. H. (1988). Internal heterogeneity of gaps and species richness in Costa Rican tropical wet forest. *Journal of Tropical Ecology*, **4**, 99–119.

Brokaw, N. V. L. & Schiener, S. M. (1989). Species composition in gaps and structure of a tropical forest. *Ecology*, **70**, 538–41.

Brooke, M. de L. (1983). Ecological segregation of woodcreepers Dendrocolaptidae in the state of Rio de Janeiro, Brazil. *Ibis*, **125**, 562–7.

Brosset, A. (1990). A long term study of the rain forest birds in M'Passa, Gabon. In *Biogeography and ecology of forest birds*, ed. A. Keast, pp. 259–74. The Hague: SPB Academic Publishing.

Brown, N. (1993). The implications of climate and gap microclimate for seedling growth conditions in a Bornean lowland rain forest. *Journal of Tropical Ecology*, **9**, 153–68.

Brown, N. D. (1990). Dipterocarp regeneration in tropical rain forest gaps of different sizes. DPhil thesis, University of Oxford.

Brown, N. D. & Whitmore, T. C. (1992). Do dipterocarp seedlings really partition tropical rain forest gaps? In *Tropical rain forest: disturbance and recovery*, ed. A. G. Marshall & M. D. Swaine, pp. 369–78. London: The Royal Society.

Brown, S. & Lugo, A. E. (1984). Biomass of tropical forests: a new estimate based on forest volumes. *Science*, **223**, 1290–3.

Bruijnzeel, L. A. (1992). Managing tropical forest watersheds for production: where contradictory theory and practice co–exist. In *Wise management of tropical forests*, ed. F. R. Miller & K. L. Adam, pp. 37–75. Oxford: Oxford Forestry Institute.

Bruijnzeel, L. A. & Critchley, W. R. S. (1994). *Environmental impacts of logging moist tropical forests*. IHP Humid Tropics Programme Series no. 7. Paris: UNESCO.

Brundrett, M. (1991). Mycorrhizas in natural ecosystems. *Advances in Ecological Research*, **21**, 171–313.

Burgess, P. F. (1971). Effect of logging on hill dipterocarp forest. *Malayan Nature Journal*, **24**, 231–7.

Burghouts, T., Ernsting, G., Korthals, G. & de Vries, T. (1992). Litterfall, leaf litter decomposition and litter invertebrates in primary and selectively logged dipterocarp forest in Sabah, Malaysia. In *Tropical rain forest: disturbance and recovery*, ed. A. G. Marshall & M. D. Swaine, pp. 407–16. London: The Royal Society.

Burgman, M., Church, R., Ferguson, I., Gijsbers, R., Lau, A., Lindenmayer, D., Loyn, R., McCarthy, M. & Vandenburgh, W. (1994). Wildlife planning using FORPLAN: a review and examples from Victoria forests. *Australian Forester*, **57**, 131–40.

Caldecott, J. (1988). *Hunting and wildlife management in Sarawak*. Gland, Switzerland: IUCN.

Canham, C. D. (1989). Different responses to gaps among shade-tolerant tree species.

*Ecology*, **70**, 548–50.

Cannon, C. H., Peart, D. R., Leighton, M. & Kartawinata, K. (1994). The structure of lowland rainforest after selective logging in West Kalimantan, Indonesia. *Forest Ecology and Management*, **67**, 49–68.

Cassell, D. S., Gilmour, D. A. & Bonell, M. (1984). Watershed forest management practices in the tropical rain forests of north-eastern Australia. In *Effects of forest land use on erosion and slope stability*, ed. C. L. O'Loughlin & A. J. Pearce, pp. 289–98. Vienna: IUFRO.

Chai, D. N. P. & Udarbe, M. P. (1977). The effectiveness of current silvicultural practice in Sabah. *Malaysian Forester*, **40**, 27–35.

Chapman, C. A. & Chapman, L. J. (1996). Exotic tree plantations and the regeneration of natural forests in Kibale National Park, Uganda. *Biological Conservation*, **76**, 253–7.

Chazdon, R. L. & Fetcher, N. (1984). Photosynthetic light environments in a tropical lowland rain forest in Costa Rica. *Journal of Ecology*, **72**, 553–64.

Chepko-Sade, B. D. & Halpin, Z. T. (1987). *Mammalian dispersal patterns*. Chicago: University of Chicago Press.

Chivers, D. J. (1972). The siamang and the gibbon in the Malay peninsula. In *Gibbon and Siamang*, vol. 1, ed. D. M. Rumbaugh, pp. 102–36. Basel: Karger.

Chivers, D. J. & Hladik, C. M. (1980). Morphology of the gastro-intestinal tract in primates: some comparisons with other mammals in relation to diet. *Journal of Morphology*, **166**, 337–86.

Christensen, B. (1978). Mangroves: what are they worth? *Unasylva*, **30**, 2–15.

Clubbe, C. P. & Jhilmit, S. (1992). A case study of natural forest management in Trinidad. In *Wise management of tropical forests*, ed. F. R. Miller & K. L. Adam, pp. 201–9. Oxford: Oxford Forestry Institute.

Coimbra-Filho, A. F. & Mittermeier, R. A. (1976). Exudate-eating and tree-gouging in marmosets. *Nature*, **262**, 630.

Collar, N. J., Crosby, M. J. & Stattersfield, A. J. (1994). *Birds to watch 2. The world list of threatened birds*. Cambridge: BirdLife International.

Collins, N. M. (1980). The effect of logging on termite (Isoptera) diversity and decomposition processes in lowland dipterocarp forests. In *Tropical ecology and development*, vol. 1, ed. J. I. Furtado, pp. 113–21. Kuala Lumpur: International Society of Tropical Ecology.

Collins, N. M., Sayer, J. A. & Whitmore, T. C. (1991). *The conservation atlas of tropical forests: Asia and the Pacific*. London: Macmillan.

Crome, F. H. J. (1991). Wildlife conservation and rain forest management: examples from North-east Queensland. In *Rain forest regeneration and management*, ed. A. Gómez-Pompa, T. C. Whitmore & M. Hadley, pp. 407–18. Paris: UNESCO.

Crome, F. H. J. & More, L. A. (1989). Display site constancy of bowerbirds and the effects of logging on Mt Windsor tableland, north Queensland. *Emu*, **89**, 47–52.

Crome, F. H. J., More, L. A. & Richards, G. C. (1992). A study of logging damage in upland rainforest in north Queensland. *Forest Ecology and Management*, **49**, 1–29.

Crome, F. H. J. & Richards, G. C. (1988). Bats and gaps: microchiropteran community structure in a Queensland rain forest. *Ecology*, **69**, 1960–9.

Danielsen, F. & Heegaard, M. (1995). Impact of logging and plantation development on species diversity: a case study from Sumatra. In *Management of tropical forests: towards an integrated perspective*, ed. O. Sandbukt, pp. 73–92. Oslo: Centre for Development and the Environment, University of Oslo.

Date, E. M., Ford, H. A. & Recher, H. F. (1991). Frugivorous pigeons, stepping stones and weeds in northern New South Wales. In *Nature conservation 2: the role of corridors*, ed. D. A. Saunders & R. J. Hobbs, pp. 241–5. Chipping Norton, Australia: Surrey Beaton and Sons.

Davidson, J. (1985). Plantation forestry in relation to tropical moist forests in South-east Asia. In *The future of tropical rain forests in South-east Asia*. IUCN Commission on Ecology Paper **10**, 101–10.

Davies, A. G. (1987). *The Gola Forest Reserves, Sierra Leone: wildlife conservation and forest management*. Gland, Switzerland: IUCN.

Davies, A. G. & Payne, J. B. (1982). *A faunal survey of Sabah*. Kuala Lumpur, Malaysia: WWF Malaysia.

Davies, G. (1986). The orang-utan in Sabah. *Oryx*, **20**, 40–5.

Dawkins, H. C. (1958). *The management of natural tropical high forest with special reference to Uganda*. Oxford: Imperial Forestry Institute.

Diamond, J. M., Bishop, K. D. & van Balen, S. (1987). Bird survival in an isolated Javan woodlot: island or mirror? *Conservation Biology*, **2**, 132–42.

Dickinson, R. E. & Kennedy, P. (1992). Impacts on regional climate of Amazon deforestation. *Geophysical Research Letters*, **19**, 1947–50.

Dittus, W. P. G. (1977). The social regulation of population density and age–sex distribution in the toque monkey. *Behaviour*, **63**, 281–322.

Dixon, J. & Sherman, P. (1990). *Economics of protected areas: a new look at benefits and costs*. Washington, DC: Island Press.

Dodson, C. H. & Gentry, A. H. (1991). Biological extinction in western Ecuador. *Annals of the Missouri Botanic Garden*, **78**, 273–95.

Douglas, I., Greer, A. J., Wong, W. M., Spencer, T. & Sinun, W. (1990). The impact of commercial logging on a small rain forest catchment in Ulu Segama, Sabah, Malaysia. *International Association of Hydrological Sciences Publication*, **75**, 17–30.

Douglas, I., Spencer, T., Greer, A. J., Kawi Bidin, Sinun, W. & Wong, W. M. (1992). The impact of selective commercial logging on stream hydrology, chemistry and sediment loads in the Ulu Segama rain forest, Sabah. In *Tropical rain forest: disturbance and recovery*, ed. A. G. Marshall and M. D. Swaine, pp. 397–406. London: The Royal Society.

Dransfield, J. (1988). Prospects for rattan cultivation. *Advances in Economic Botany*, **6**, 190–200.

Dranzoa, C. (1990). Survival of birds in formerly forested areas around Kampala. MSc thesis, Makerere University, Kampala, Uganda.

Dranzoa, C. (1995). Effects of forest management and the role of small patches on bird communities of the Kibale Forest National Park, Uganda. PhD thesis, Makerere University, Kampala.

Dranzoa, C. & Johns, A. D. (1992). Recovery of bird populations in intensively managed forest in southern Uganda. *WCI-Uganda Miscellaneous Research Reports*, **1**, 1–38.

Driscoll, P. V. & Kikkawa, J. (1989). Bird species diversity of lowland tropical rainforests of New Guinea and northern Australia. In *Vertebrates in complex tropical systems*, ed. M. L. Harmelin-Vivien & F. Bourliere, pp. 123–52. Berlin, Heidelberg, New York: Springer.

Duff, A. B., Hall, R. A. & Marsh, C. W. (1984). A survey of wildlife in and around a commercial tree plantation in Sabah. *Malaysian Forester*, **47**, 197–13.

Du Plessis, M. A. (1995). The effects of fuelwood removal on the diversity of some cavity-using birds and mammals in South Africa. *Biological Conservation*, **74**, 77–82.

Earl, D. E. (1968). *Latest techniques in the treatment of natural high forest in south Mengo district*. Entebbe: Government Printer.

Earl, D. E. (1973). Does forestry need a new ethos? *Commonwealth Forestry Review*, **151**, 82–9.

Earl, D. E. (1992). Wise management of natural tropical forest for timber production, tourism and wildlife. In *Wise management of tropical forests*, ed. F. R. Miller & K. L.

Adam, pp. 211–19. Oxford: Oxford Forestry Institute.

Eggleton, P., Bignell, D. E., Sands, W. A., Waite, B., Wood, T. G. & Lawton, J. H. (1995). The species richness of termites (Isoptera) under differing levels of forest disturbance in the Mbalmayo Forest Reserve, southern Cameroon. *Journal of Tropical Ecology*, **11**, 85–98.

Emmons, L. H. (1980). Ecology and resource partitioning among nine species of African rain forest squirrels. *Ecological Monographs*, **50**, 31–54.

Emmons, L. H., Gautier-Hion, A. & Dubost, G. (1983). Community structure of the frugivorous-folivorous forest mammals of Gabon. *Journal of Zoology, London*, **199**, 209–22.

Ewel, J. & Conde, L. (1976). Potential ecological impact of increased intensity of tropical forest utilization. Final report to USAID Research Agreement 12–28. Forest Products Laboratory, U.S.D.A. Forest Service, Madison, Wisconsin.

FAO (1981). *Tropical forest resources assessment project*, 4 vols. Rome: FAO.

FAO (1988). *An interim report on the state of forest resources in the developing countries.* Rome: FAO.

FAO (1989). *Review of forest management systems of tropical Asia.* Rome: FAO.

FAO (1991a). *Second interim report on the state of tropical forests: forest resources assessment 1990 project.* Rome: FAO.

FAO (1991b). *Exotic aphid pests of conifers: a crisis in African forestry. Workshop proceedings, Kenya Forestry Research Institute.* Rome: FAO.

FAO (1992). *The forest resources of the tropical zone by main ecological regions: forest resources assessment 1990 project.* Rome: FAO.

FAO (1993). *Assessing forestry project impacts: issues and strategies.* Rome: FAO.

Fearnside, P. (1989). Forest management in Amazonia: the needs for new criteria in evaluating development options. *Forest Ecology and Management*, **27**, 61–79.

Feinsinger, P. (1976). Organization of a tropical guild of nectarivorous birds. *Ecological Monographs*, **46**, 257–91.

Fetcher, N., Oberbauer, S. F. and Strain, B. R. (1984). Vegetation effects on microclimate in lowland tropical forest in Costa Rica. *International Journal of Biometeorology*, **29**, 145–55.

Fimbel, C. (1994). Ecological correlates of species success in modified habitats may be disturbance- and site-specific: the primates of Tiwai Island. *Conservation Biology*, **8**, 106–13.

Fogden, M. P. L. (1972). The seasonality and population dynamics of equatorial forest birds in Sarawak. *Ibis*, **114**, 307–43.

Forman, R. T. T. (1991). Landscape corridors: from theoretical foundations to public policy. In *Nature conservation 2: the role of corridors*, ed. D. A. Saunders & R. J. Hobbs, pp. 71–84. Chipping Norton, Australia: Surrey Beaton and Sons.

Foster, S. A. & Janson, C. H. (1985). The relationship between seed size and establishment conditions in tropical woody plants. *Ecology*, **66**, 773–80.

Fox, J. E. D. (1968). Logging damage and the influence of climber cutting in the lowland dipterocarp forests in Sabah. *Malayan Forester*, **3**, 326–47.

Franklin, J. F. (1989). Toward a new forestry. *American Forests*, **95**, 37–44.

Franklin, J. F. & Forman, R. T. T. (1987). Creating landscape patterns by forest cutting: ecological consequences and principles. *Landscape Ecology*, **1**, 5–18.

Freezaillah B. C. Y. (1984). Lesser-known tropical wood species: how bright is their future. *Unasylva*, **145**, 3–16.

Frith, C. B. & Frith, D. W. (1985). Seasonality of insect abundance in an Australian upland tropical rainforest. *Australian Journal of Ecology*, **10**, 237–48.

Galetti, M. & Chivers, D. J. (1995). Palm harvest threatens Brazil's best protected area of Atlantic forest. *Oryx*, **29**, 225–26.

Ganzhorn, J. U., Ganzhorn, A. W., Abraham, J-P., Andriamanarivo, L. & Ramananjatovo, A. (1990). The impact of selective logging on forest structure and tenrec populations in western Madagascar. *Oecologia*, **84**, 126–33.

Gautier-Hion, A. & Michaloud, G. (1989). Are figs always keystone resources for tropical frugivorous vertebrates? A test in Gabon. *Ecology*, **70**, 1826–33.

Gentry, A. H. (1986). Endemism in tropical versus temperate plant communities. In *Conservation Biology*, ed. M. E. Soulé, pp. 153–81. Sunderland, Mass.: Sinauer.

Geollogue, R. T. & Hue, R. (1981). Early stages of forest regeneration in south Sumatra. *BIOTROP Special Publications*, **13**, 153–61.

Gilbert, L. E. (1980). Food web organization and the conservation of Neotropical diversity. In *Conservation Biology*, ed. M. E. Soulé, pp. 11–33. Sunderland, Mass.: Sinauer.

Gill, A. E. & Rasmussen, E. M. (1983). The 1982–83 climate anomaly in the equatorial Pacific. *Nature*, **306**, 229–34.

Gillis, M. (1988). Indonesia: public policies, resource management and the tropical forest. In *Public policies and the misuse of forest resources*, ed. R. Repetto & M. Gillis, pp. 43–113. Cambridge: Cambridge University Press.

Gillman, G. P., Sinclair, D. F., Knowlton, R. & Keys, M. G. (1985). The effect on some soil chemical properties of the selective logging of a north Queensland rain forest. *Forest Ecology and Management*, **12**, 195–214.

Gómez-Pompa, A. & Burley, F. W. (1991). The management of natural tropical forests. In *Rain forest regeneration and management*, ed. A. Gómez-Pompa, T. C. Whitmore & M. Hadley, pp. 3–18. Paris: UNESCO.

Gómez-Pompa, A., Flores, S. & Sosa, V. (1987). The 'pet kot': a man-made tropical forest of the Maya. *Interciencia*, **12**, 10–15.

Graaf, N. R. de (1986). *A silvicultural system for natural regeneration of tropical rain forest in Suriname*. Wageningen, The Netherlands: Pudoc.

Grajal, A. (1995). Digestive efficiency of the hoatzin, *Opisthocomus hoazin*, a folivorous bird with foregut fermentation. *Ibis*, **137**, 383–8.

Grieser Johns, A. (1996). Bird population persistence in Sabahan logging concessions. *Biological Conservation*, **75**, 3–10.

Grieser Johns, A. & Grieser Johns, B. (1995). Tropical forest primates and logging: long-term co-existence? *Oryx*, **29**, 205–11.

Hamilton, L. S. (1985). Some water-soil consequences of modifying tropical rain forests. In *The future of tropical rain forests in South-east Asia*. IUCN Commission on Ecology Paper, **10**, 69–80.

Harcourt, C. & Thornback, J. (1990). *Lemurs of Madagascar and the Comoros: the IUCN Red Data Book*. Gland, Switzerland and Cambridge: IUCN.

Hartshorn, G. S. (1990). Natural forest management by the Yanesha forestry cooperative in Peruvian Amazonia. In *Alternatives to deforestation: steps towards sustainable use of the Amazon rain forest*, ed. A. B. Anderson, pp. 128–38. New York: Columbia University Press.

Harris, L. D. (1984). *The fragmented forest*. Chicago, Illinois: University of Chicago Press.

Harrison, R. L. (1992). Towards a theory of inter-refuge corridor design. *Conservation Biology*, **6**, 293–5.

Hawkins, A. F. A., Chapman, P., Ganzhorn, J. U., Bloxham, Q. M. C., Barlow, S. C. & Tonge, S. J. (1990). Vertebrate conservation in Ankarana and Analamera Special Reserves, northern Madagascar. *Biological Conservation*, **54**, 83–110.

Hawthorne, W. D. (1993). Forest regeneration after logging: findings of a study in the Bia

South Game Production Reserve, Ghana. *ODA Forestry Series*, **3**.

Henderson-Sellers, A., Dickinson, R. E., Durbridge, T. B., Kennedy, P. J., McGuffie, K. & Pitman, A. J. (1993). Tropical deforestation: modelling local- to regional-scale climate change. *Journal of Geophysical Research*, **98**, 7289–315.

Hendrison, J. 1990. *Damage-controlled logging in managed tropical rain forest in Suriname*. Wageningen, The Netherlands: Pudoc.

Henwood, A. (1986). Moth trapping in the rain forest of Borneo. Report to Yayasan Sabah, Kota Kinabalu.

Heydon, M. J. (1994). The ecology and management of rain-forest ungulates in Sabah, Malaysia: implications of forest disturbance. Report to ODA/NERC project F3CR26/G1/05, Institute of South-east Asian Biology, University of Aberdeen.

Heydon, M. J. & Bulloh, P. (1996). The impact of selective logging on sympatric civet species in Borneo. *Oryx*, **30**, 31–6.

Heywood, V. H. & Stuart, S. N. (1992). Species extinctions in tropical forests. In *Tropical deforestation and species extinction*, ed. T. C. Whitmore & J. A. Sayer, pp. 91–117. London: Chapman and Hall.

HIID (1988). *The case for multiple use management of tropical hardwood forests*. Cambridge, Mass.: Harvard Institute for International Development.

Hladik, C. M., Charles-Dominique, P. & Petter, J-J. (1980). Feeding strategies of five nocturnal prosimians in the dry forests of the west coast of Madagascar. In *Nocturnal Malagasy Primates*, ed. P. Charles-Dominique, H. M. Cooper, A. Hladik, C. M. Hladik, E. Pages, G. F. Pariete, A. Petter-Rousseau, J-J. Petter & A. Schilling, pp. 41–73. New York: Academic Press.

Holbech, L. H. (1992). Effects of selective logging on a rain forest bird community in western Ghana. MSc thesis, University of Copenhagen.

Holloway, J. D. (1984a). Notes on the butterflies of the Gunung Mulu National Park. *Sarawak Museum Journal*, 30, Special Issue 2.

Holloway, J. D. (1984b). Moths as indicator organisms for categorizing rain-forest and monitoring changes and regeneration processes. In *Tropical rain-forest: the Leeds symposium*, ed. A. C. Chadwick & S. L. Sutton, pp. 235–42. Leeds: Leeds Philosophical and Literary Society.

Holloway, J. D. & Herbert, P. D. N. (1979). Ecological and taxonomic trends in macrolepidopteran host plant selection. *Biological Journal of the Linnaean Society*, **11**, 229–51.

Holloway, J. D., Kirk-Spriggs, A. H. & Chey, V. K. (1992). The response of some rain forest insect groups to logging and conversion to plantations. In *Tropical rain forest: disturbance and recovery*, ed. A. G. Marshall & M. D. Swaine, pp. 425–36. London: The Royal Society.

Hopkins, A. J. M. & Saunders, D. A. (1987). Ecological studies as the basis for management. In *Nature Conservation: the role of remnants of native vegetation*, ed. D. A. Saunders, G. W. Arnold, A. A. Burbridge & A. J. M. Hopkins, pp. 15–28. Chipping Norton, Australia: Surrey Beaton and Sons.

Horn, H. S. (1966). Measurement of overlap in comparative ecological studies. *American Naturalist*, **100**, 419–24.

Howard, P. C. (1991). *Nature conservation in Uganda's tropical forest reserves*. Gland, Switzerland: IUCN.

Howard, W. J., Ayres, J. M., Lima-Ayres, D. & Armstrong, G. (1995). Mamirauá: a case study of biodiversity conservation involving local people. *Commonwealth Forestry Review*, **74**, 76–9.

Hunter, M. L. Jr. (1990). *Wildlife, forests and forestry*. Englewood Cliffs, New Jersey:

Prentice-Hall.

IPCC (Intergovernmental Panel on Climate Change) (1990). *Climate change: the IPCC scientific assessment.* Cambridge: Cambridge University Press.

IPCC (Intergovernmental Panel on Climate Change) (1992). *Climate change 1992: the supplementary report to the IPCC scientific assessment.* Cambridge: Cambridge University Press.

IPCC (Intergovernmental Panel on Climate Change) (1994). *Climate change 1994: radiative forcing of climate change and an evaluation of the IPCC IS92 emissions scenarios.* Cambridge: Cambridge University Press.

Irmler, U. (1979). Abundance fluctuations and habitat changes of soil beetles in central American inundation forests (Coleoptera: Carabidae, Staphylinidae). *Studies of Neotropical Fauna and Environment,* **14,** 1–16.

ITTO (1989). *Pre-project study report: enrichment planting.* Yokohama, Japan: International Tropical Timber Organization.

ITTO (1990). *Guidelines for sustainable management of natural tropical forests.* Yokohama, Japan: International Tropical Timber Organization.

IUCN (1986). *IUCN directory of Afrotropical protected areas.* Gland, Switzerland: IUCN

IUCN/UNEP/WWF (1980). *World conservation strategy.* Gland, Switzerland: IUCN.

IUCN/UNEP/WWF (1991). *Caring for the earth: a strategy for sustainable living.* Gland, Switzerland: IUCN.

Isabirye-Basuta, G. & Kasenene, J. M. (1987). Small rodent populations in selectively felled and mature tracts of Kibale Forest, Uganda. *Biotropica,* **19,** 260–6.

Janos, D. P. (1980). Mycorrhizae influence tropical succession. *Biotropica,* 12 (suppl.), 56–64.

Johns, A. D. (1983). Ecological effects of selective logging in a West Malaysian rain forest. PhD thesis, University of Cambridge.

Johns, A. D. (1985). Selective logging and wildlife conservation in tropical rain-forest: problems and recommendations. *Biological Conservation,* **31,** 355–75.

Johns, A. D. (1986a). Effects of habitat disturbance on rain forest wildlife in Brazilian Amazonia. Final report to project US-302, World Wildlife Fund US, Washington DC.

Johns, A. D. (1986b). Effects of selective logging on the behavioral ecology of West Malaysian primates. *Ecology,* **67,** 684–94.

Johns, A. D. (1986c). Effects of selective logging on the ecological organization of a peninsular Malaysian rain forest avifauna. *Forktail,* **1,** 65–79.

Johns, A. D. (1987). The use of primary and selectively logged rain forest by Malaysian hornbills (Bucerotidae) and implications for their conservation. *Biological Conservation,* **40,** 179–90.

Johns, A. D. (1988). Effects of 'selective' timber extraction on rain forest structure and composition and some consequences for frugivores and folivores. *Biotropica,* **20,** 31–7.

Johns, A. D. (1989a). Timber, the environment and wildlife in Malaysian rain forests. Report to ODA/NERC project F3CR26/G1/05, Institute of South-east Asian Biology, University of Aberdeen.

Johns, A. D. (1989b). Recovery of a peninsular Malaysian rain forest avifauna following selective timber logging: the first twelve years. *Forktail,* **4,** 89–105.

Johns, A. D. (1991a). Forest disturbance and Amazonian primates. In *Primate responses to environmental change,* ed. H. O. Box, pp. 115–35. London: Chapman and Hall.

Johns, A. D. (1991b). Responses of Amazonian rain forest birds to habitat modification. *Journal of Tropical Ecology,* **7,** 417–37.

Johns, A. D. (1992). Species conservation in managed tropical forests. In *Tropical deforestation and species extinction,* ed. T. C. Whitmore & J. A. Sayer, pp. 15–53. London: Chapman and Hall.

Johns, A. D. & Skorupa, J. P. (1987). Responses of rain forest primates to habitat disturbance: a review. *International Journal of Primatology*, **8**, 157–91.

Johns, A. D., Pine, R. J. & Wilson, D. E. (1985). Rain forest bats: an uncertain future. *Bat News*, **5**, 4–5.

Johns, R. (1992). The influence of deforestation and selective logging operations on plant diversity in Papua New Guinea. In *Tropical deforestation and species extinction*, ed. T. C. Whitmore & J. A. Sayer, pp. 143–7. London: Chapman and Hall.

Jones, C. B. (1994). Injury and disease of the mantled howler monkey in fragmented habitats. *Neotropical Primates*, **2**(4), 4–5.

Jonkers, W. B. J. (1987). *Vegetation structure, logging damage and silviculture in a tropical rain forest in Suriname.* Wageningen, The Netherlands: Agricultural University.

Jonsson, T. & Lindgren, P. (1990). *Logging technology for tropical forests: for or against?* Krista, Sweden: Forest Operations Institute.

Kalina, J. (1988). Ecology and behaviour of the black–and–white casqued hornbill (*Bycanistes subcylindricus subquadratus*) in Kibale Forest, Uganda. PhD thesis, Michigan State University.

Kalyakin, M. V., Korzun, L. P. & Trunov, V. I. (1994). On the biodiversity of birds in lowland tropical forests of south Vietnam. In *International Symposium on Biodiversity and Systematics in Tropical Ecosystems, Abstracts*, pp. 37–8. Zoologisches Forschungsinstitut und Museum Alexander Koenig, Bonn, Germany.

Kamaruzaman, J. (1991). A survey of soil disturbance from tractor logging in a hill forest of peninsular Malaysia. In *Malaysian forestry and forest products research*, ed. S. Appanah, F. S. P. Ng & R. Ismail, pp. 16–21. Kepong, Malaysia: Forest Research Institute of Malaysia.

Kanowski, P. J., Savill, P. S., Adlard, P. G., Burley, J., Evans, J., Palmer, J. R. & Wood, P. J. (1992). Plantation forestry. In *Managing the world's forests*, ed. N. P. Sharma, pp. 375–401. Dubuque, Iowa: Kendall/Hunt Publishing.

Kapos, V. (1989). Effects of isolation on the water status of forest patches in the Brazilian Amazon. *Journal of Tropical Ecology*, **5**, 173–85.

Karr, J. R. (1976). Seasonality, resource availability and community diversity in Neotropical bird communities. *American Naturalist*, **110**, 973–94.

Karr, J. R. (1980). Geographical variation in the avifaunas of tropical forest undergrowth. *Auk*, **97**, 283–98.

Karr, J. R. (1982). Population variability and extinction in the avifauna of a tropical land-bridge island. *Ecology*, **63**, 1975–8.

Karr, J. R. (1990). Birds of tropical rainforest: comparative biogeography and ecology. In *Biogeography and ecology of forest bird communities*, ed. A. Keast, pp. 215–28. The Hague: SPB Academic Publishing.

Karr, J. R. & Freemark, K. E. (1983). Habitat selection and environmental gradients: dynamics in the 'stable' tropics. *Ecology*, **64**, 1481–94.

Karr, J. R. & Roth, R. R. (1971). Vegetation structure and avian diversity in several New World areas. *American Naturalist*, **105**, 423–35.

Kasenene, J. M. (1987). The influence of mechanized selective logging, felling intensity and gap–size on the regeneration of a tropical moist forest in the Kibale Forest Reserve, Uganda. PhD thesis, Michigan State University.

Kasenene, J. M. & Murphy, P. G. (1991). Post-logging tree mortality and major branch losses in Kibale Forest, Uganda. *Forest Ecology and Management*, **46**, 295–307.

Kavanagh, R. P. & Bamkin, K. L. (1995). Distribution of nocturnal forest birds and mammals in relation to the logging mosaic in south-eastern New South Wales, Australia. *Biological Conservation*, **71**, 41–53.

Kemp, A. C. ed. (1985). ICBP hornbill specialist group communication, 5. Cambridge: International Council for Bird Preservation.

Kemp, A. C. (1995). *The hornbills.* Oxford: Oxford University Press.

Kemp, A. C. & Kemp, M. I. (1975). Report on a study of hornbills in Sarawak, with comments on their conservation. Final report to project 2/74, World Wildlife Fund Malaysia, Kuala Lumpur.

Kennedy, D. N. & Swaine, M. D. (1992). Germination and growth of colonizing species in artificial gaps of different sizes in dipterocarp rain forest. In *Tropical rain forest: disturbance and recovery,* ed. A. G. Marshall & M. D. Swaine, pp. 357–66. London: The Royal Society.

Khan, N. (1995). Protection of north Selangor peat swamp forest, Malaysia. *Parks,* **5**(2), 24–31.

Kio, P. R. O. & Ekwebelam, S. A. (1987). Plantations versus natural forests for meeting Nigeria's wood needs. In *Natural management of tropical moist forests,* ed. F. Mergen & J. Vincent, pp. 149–76. New Haven: Yale University Press.

Kleine, M. & Heuveldop, J. (1993). A management planning concept for sustained yield of tropical forests in Sabah, Malaysia. *Forest Ecology and Management,* **61**, 277–97.

Korpelainen, H., Adjers, G., Kuusipalo, J., Nuryanto, K. & Otsamo, A. (1995). Profitability of rehabilitation of overlogged dipterocarp forest: a case study from South Kalimantan, Indonesia. *Forest Ecology and Management,* **79**, 207–15.

Kremen, C., Merenlender, A. M. & Murphy, D. D. (1994). Ecological monitoring: a vital need for integrated conservation and development programmes in the tropics. *Conservation Biology,* **8**, 388–97.

Laidlaw, R. K. (1994). The Virgin Jungle Reserves of peninsular Malaysia: the ecology and dynamics of small protected areas in managed forest. PhD thesis, University of Cambridge.

Lamb, D. (1990). *Exploiting the tropical rain forest: an account of pulpwood logging in Papua New Guinea.* Paris: UNESCO.

Lambert, F. R. (1990). Avifaunal changes following selective logging of a north Bornean rain forest. Report to ODA/NERC project F3CR26/G1/05, Institute of Tropical Biology, University of Aberdeen.

Lambert, F. R. (1992). The consequences of selective logging for Bornean lowland forest birds. In *Tropical rain forest: disturbance and recovery,* ed. A. G. Marshall & M. D. Swaine, pp. 443–57. London: The Royal Society.

Lambert, F. R. & Marshall, A. G. (1991). Keystone characteristics of bird-dispersed *Ficus* in a Malaysian lowland rain forest. *Journal of Ecology,* **79**, 793–809.

Landres, P. B., Verner, J. & Thomas, J. W. (1988). Ecological uses of vertebrate indicator species: a critique. *Conservation Biology,* **2**, 316–28.

Laurance, W. F. (1991). Ecological correlates of extinction proneness in Australian tropical rain forest mammals. *Conservation Biology,* **5**, 79–89.

Laurance, W. F. & Laurance, S. G. W. (1996). Responses of five arboreal marsupials to recent selective logging in tropical Australia. *Biotropica,* 28, 310–22.

Lavery, H. J. & Grimes, R. J. (1974). The function of the bill in the tooth-billed bowerbird. *Emu,* **74**, 974–5.

Lean, J. & Rowntree, P. R. (1993). A GCM simulation of the impact of Amazonian deforestation on climate using an improved canopy representation. *Quarterly Journal of the Royal Meteorological Society,* **119**, 509–30.

Lee, D. W. (1989). Canopy dynamics and light climates in a tropical moist deciduous forest in India. *Journal of Tropical Ecology,* **5**, 65–79.

Lee, H. S. (1982). The development of silvicultural systems in the hill forests of Malaysia.

*Malaysian Forester,* **45**, 1–9.

Leighton, M. & Leighton, D. R. (1983). Vertebrate responses to fruiting seasonality within a Bornean rain forest. In *Tropical rain forest: ecology and management,* ed. S. L. Sutton, T. C. Whitmore & A. C. Chadwick, pp. 181–96. Oxford: Blackwell Scientific Publications.

Leighton, M. & Wirawan, N. (1986). Catastrophic drought and fire in Borneo tropical rain forest associated with the 1982–83 El Niño Southern Oscillation event. In *Tropical rain forests and the world atmosphere,* ed. G. T. Prance, pp. 75–102. Boulder, Colorado: Westview Press.

Leopold, A. (1949). *A Sand County almanac.* New York: Oxford University Press.

Leslie, A. J. (1987). A second look at the economics of natural management systems in tropical mixed forests. *Unasylva,* **155**, 46–58.

Levings, S. C. & Windsor, D. M. (1983). Seasonal and annual variation in litter arthropod populations. In *The ecology of a tropical rain forest: seasonal rhythms and long-term changes,* ed. E. G. Leigh, A. S. Rand and D. M. Windsor, pp. 355–87. Oxford: Oxford University Press.

Levins, R. (1968). *Evolution in changing environments.* Princeton, New Jersey: Princeton University Press.

Lim, B. H. & Sasekumar, A. (1979). A preliminary study on the feeding biology of mangrove forest primates, Kuala Selangor. *Malaysian Nature Journal,* **33**, 105–12.

Lindberg, K. & Huber, R. M. Jr. (1993). Economic issues in ecotourism management. In *Ecotourism: a guide for planners and managers,* ed. K. Lindberg & D. E. Hawkins, pp. 82–115. North Bennington, Vermont: The Ecotourism Society.

Lindenmayer, D. B. & Nix, H. A. (1993). Ecological principles for the design of wildlife corridors. *Conservation Biology,* **7**, 627–30.

Lindenmayer, D. B., Cunningham, R. B., Tanton, M. T., Smith, A. P. & Nix, H. A. (1990a). The conservation of arboreal marsupials in the montane ash forests of the Central Highlands of Victoria, south-east Australia, I. Factors influencing the occupancy of trees with hollows. *Biological Conservation,* **54**, 111–31.

Lindenmayer, D. B., Cunningham, R. B., Tanton, M. T. & Smith, A. P. (1990b). The conservation of arboreal marsupials in the montane ash forests of Victoria, south-east Australia, II. The loss of trees with hollows and its implications for the conservation of Leadbeater's possum *Gymnobelideus leadbeateri* McCoy (Marsupialia: Petauridae). *Biological Conservation,* **54**, 133–45.

Lindenmayer, D. B., Cunningham, R. B., Tanton, M. T., Nix, H. A. & Smith, A. P. (1991). The conservation of arboreal marsupials in the montane ash forests of the Central Highlands of Victoria, south-east Australia, III. The habitat requirements of Leadbeater's possum *Gymnobelideus leadbeateri* and models of the diversity and abundance of arboreal marsupials. *Biological Conservation,* **56**, 295–315.

Lindenmayer, D. B., Cunningham, R. B. & Donnelly, C. F. (1993). The conservation of arboreal marsupials in the montane ash forests of the Central Highlands of Victoria, south-east Australia, IV. The presence and abundance of arboreal marsupials in retained linear habitats (wildlife corridors) within logged forest. *Biological Conservation,* **66**, 207–21.

Longman, K. A. & Jenik, J. (1974). *Tropical forest and its environment.* Thetford, Norfolk, UK: Lowe and Brydone.

Lovejoy, T. E. (1974). Bird diversity and abundance in Amazon forest communities. *Living bird,* **13**, 127–91.

Lovejoy, T. E., Bierregaard, R. O. Jr, Rylands, A. B., Malcolm, J. R., Quintela, C. E., Harper, L. H., Brown, K. S. R. Jr, Powell, A. H., Powell, G. V. N., Schubart, H. O. R. & Hays, M. B. (1986). Edge and other effects of isolation on Amazon forest fragments. In

*Conservation biology*, ed. M. E. Soulé, pp. 257–85. Sunderland, Mass.: Sinauer.

Lovejoy, T. E., Rankin, J. M., Bierregaard, R. O. Jr, Brown, K. S. Jr, Emmons, L. H. & van der Voort, M. E. (1984). Ecosystem decay in Amazonian forest remnants. In *Extinctions*, ed. M. H. Nitecki, pp. 295–325. Chicago: University of Chicago Press.

Lowe, R. G. (1978). Experience with the shelterwood system of regeneration in natural forest in Nigeria. *Forest Ecology and Management*, 1, 193–212.

Lynch, J. F. & Saunders, D. A. (1991). Responses of bird species to habitat fragmentation in the wheatbelt of Western Australia: interiors, edges and corridors. In *Nature conservation 2: the role of corridors*, ed. D. A. Saunders & R. J. Hobbs, pp. 143–58. Chipping Norton, Australia: Surrey Beaton and Sons.

Maítre, H. F. (1987). Natural forest management in Côte d'Ivoire. *Unasylva*, 157/158, 53–60.

Malvas, J. D. Jr. (1987). Development of forest sector planning, Malaysia. A report on the logging demonstration cum training coupe. UNDP/FAO Field Document MAL/85/004, no. 7.

Mann, L. K., Johnson, D. W., West, D. C., Cole, D. W., Hornbeck, J. W., Martin, C. W., Riekerk, H., Smith, C. T., Swank, W. T., Tritton, L. M. & van Lear, D. H. (1988). Effect of whole-tree and stem-only clearcutting on post harvest hydrologic losses, nutrient capital and regrowth. *Forest Science*, 34, 412–28.

Marn, H. M. (1982). The planning and design of the forest harvesting and log transport operation in the mixed dipterocarp forest of Sarawak. UNDP/FAO Field Document MAL/76/008, no. 17.

Marrs, R. H., Thompson, J., Scott, D. & Proctor, J. (1991). Nitrogen mineralization and nutrification in terra firme forest and savanna soils on Ihla de Maracá, Roraima, Brazil. *Journal of Tropical Ecology*, 7, 123–37.

Marsh, C. W. (1993). Carbon dioxide offsets as potential funding for improved tropical forest management. *Oryx*, 27, 2–3.

Marsh, C. W. & Greer, A. G. (1992). Forest land-use in Sabah, Malaysia: an introduction to Danum Valley. In *Tropical rain forest: disturbance and recovery*, ed. A. G. Marshall & M. D. Swaine, pp. 331–9. London: The Royal Society.

Marsh, C. W. & Wilson, W. L. (1981). *A survey of primates in peninsular Malaysian forests*. Kuala Lumpur: Universiti Kebangsaan Malaysia.

Marsh, C. W., Johns, A. D. & Ayres, J. M. (1987). Effects of habitat disturbance on rain forest primates. In *Primate conservation in the tropical rain forest*, ed. C. W. Marsh & R. A. Mittermeier, pp. 83–107. New York: Alan R. Liss.

Mascarenhas, B. M. (1985). 'Operação Curupira: relatório téchnico. Brasília: Eletronorte.

McCarthy, M. A. & Burgman, M. A. (1995). Coping with uncertainty in forest wildlife planning. *Forest Ecology and Management*, 74, 23–36.

McClure, H. E. (1968). Some problems concerning endangered birds in south-eastern Asia. *IUCN Publications (New Series)*, 10, 307–11.

McKey, D. & Waterman, P. G. (1982). Ranging behaviour of a group of black colobus (*Colobus satanas*) in the Douala-Edea Reserve, Cameroon. *Folia Primatologica*, 39, 264–304.

Medway, Lord & Wells, D. R. (1971). Diversity and density of birds and mammals at Kuala Lompat, Pahang. *Malayan Nature Journal*, 24, 238–47.

Meiggs, R. (1982). *Trees and timber in the ancient Mediterranean world*. Oxford: Clarendon Press.

Meijer, W. (1970). Regeneration of tropical lowland forest in Sabah, Malaysia, forty years after logging. *Malaysian Forester*, 33, 204–29.

Mendoza, G. A. & Ayemou, A. O. (1992). Analysis of some forest management strategies in Côte d'Ivoire: a regional case study. *Forest Ecology and Management*, 47, 149–74.

Menkhorst, P. W. (1984). Use of nest boxes by vertebrates in Gippsland: acceptance, preference and demand. *Australian Wildlife Research*, **11**, 255–64.

Mergen, F. & Vincent, J. R., eds. (1987). *Natural management of tropical moist forests*. New Haven: Yale University Press.

Merz, G. (1986). Movement patterns and group size of the African forest elephant *Loxodonta africana cyclotis* in the Tai National Park, Ivory Coast. *African Journal of Ecology*, **24**, 133–6.

Miller, T. B. (1981). Growth and yields of logged–over mixed dipterocarp forest in East Kalimantan. *Malaysian Forester*, **44**, 419–24.

Mitra, S. S. & Sheldon, F. H. (1993). Use of an exotic tree plantation by Bornean lowland forest birds. *Auk*, **110**, 529–40.

Mok S. T. (1992). Potential for sustainable tropical forest management in Malaysia. *Unasylva*, **169**, 28–33.

Montgomery, G. G. & Sunquist, M. E. (1978). Habitat selection and use by two-toed and three-toed sloths. In *The ecology of arboreal folivores*, ed. G. G. Montgomery, pp. 329–59. Washington DC: Smithsonian Institution Press.

Moran, G. F. & Hopper, S. D. (1987). Conservation of the genetic resources of rare and widespread eucalypts in remnant vegetation. In *Nature conservation: the role of remnants of native vegetation*, ed. D. A. Saunders, G. W. Arnold, A. A. Burbridge & A. J. M. Hopkins, pp. 151–62. Chipping Norton, Australia: Surrey Beaton and Sons.

Morton, E. S. (1978). Avian arboreal folivores: why not? In *The ecology of arboreal folivores*, ed. G. G. Montgomery, pp. 123–130. Washington, DC: Smithsonian Institution Press.

Moulds, F. R. (1988). Forest management in Australia. *Commonwealth Forestry Review*, **67**, 65–70.

Muganga, J. L. L. (1989). Population dynamics and micro–distribution of small mammals in the Kibale Forest reserve, Uganda. MSc dissertation, Makerere University, Kampala.

Munang, M. (1987). Deforestation and logging. In *Environmental conservation in Sabah*, ed. S. Sani, pp. 31–40. Kota Kinabalu, Sabah, Malaysia: Institute for Development Studies.

Muul, I. & Lim, B. L. (1978). Comparative morphology, food habits and ecology of some Malaysian arboreal rodents. In *The ecology of arboreal folivores*, ed. G. G. Montgomery, pp. 361–368. Washington, DC: Smithsonian Institution Press.

Myers, N. (1989). *Deforestation rates in tropical countries and their climatic implications*. London: Friends of the Earth.

Nair, C. T. S. (1991). A comparative account of silviculture in the tropical wet evergreen forests of Kerala, Andaman Islands and Assam. In *Rain forest regeneration and management*, ed. A. Gómez-Pompa, T. C. Whitmore & M. Hadley, pp. 371–82. Paris: UNESCO.

Nair, P. K. R. (1992). Agroforestry systems design: an ecozone approach. In *Managing the worlds forests*, ed. N. P. Sharma, pp. 375–401. Dubuque, Iowa: Kendall/Hunt Publishing.

Neil, P. E. (1984). Climber problems in Solomon Inslands forestry. *Commonwealth Forestry Review*, **63**, 27–34.

Newton, I. (1994). The role of nest-sites in limiting the numbers of hole-nesting birds: a review. *Biological Conservation*, **70**, 265–76.

Nicholson, D. I. (1958). An analysis of logging damage in tropical rain forest, North Borneo. *Malayan Forester*, **21**, 235–45.

Nicholson, D. I. (1963). Damage from high-lead logging in Sabah. *Malaysian Forester*, **26**, 294–6.

Nicholson, D. I., Henry, N. B. & Rudder, J. (1988). Stand changes in north Queensland rain forests. *Proceedings of the Ecological Society of Australia*, **15**, 61–80.

Nobre, C. A., Sellers, P. J. & Shukla, J. (1991). Amazonian deforestation and regional climate change. *Journal of Climatology*, **4**, 957–88.

Noss, R. F. (1993). Sustainable forestry or sustainable forests? In *Defining Sustainable Forestry*, ed. G. H. Aplet, N. Johnson, J. T. Olson & V. Alaric Sample, pp. 17–43. Washington, DC: Island Press.

Noss, R. F. & Harris, L. D. (1986). Nodes, networks and MUMs: preserving diversity at all scales. *Environmental Management*, **10**, 299–309.

Nummelin, M. (1989). Seasonality and effects of forestry practices on forest floor arthropods in the Kibale Forest, Uganda. *Fauna Norvegica Series B*, **36**, 17–25.

Nummelin, M. (1990). Relative habitat use of duikers, bush pigs and elephants in virgin and selectively logged areas of the Kibale Forest, Uganda. *Tropical Zoology*, **3**, 111–20.

Nussbaum, R. E. (1995). The effect of selective logging on rainforest soil and the implications for recovery. PhD thesis, University of Exeter.

Nussbaum, R., Anderson, J. & Spencer, T. (1995). Factors limiting the growth of indigenous tree seedlings planted on degraded rainforest soils in Sabah, Malaysia. *Forest Ecology and Management*, **74**, 149–59.

ODA (1995). Malaysia – United Kingdom: conservation, management and development of forest resources. Programme document, South-east Asia Development Division, Overseas Development Administration.

O'Donnell, C. F. J. (1991). Application of the wildlife corridors concept to temperate rainforest sites, North Westland, New Zealand. In *Nature conservation 2: the role of corridors*, ed. D. A. Saunders & R. J. Hobbs, pp. 85–98. Chipping Norton, Australia: Surrey Beaton and Sons.

OFI (1991). Incentives in producer and consumer countries to promote sustainable development of tropical forests. Final report to pre-project no. PCM PCF PCI(IV)/1, ITTO, Yokohama.

Oldfield, S. (1988). *Buffer zone management in tropical moist forests: case studies and guidelines*. Gland, Switzerland: IUCN.

Olivier, R. C. D. (1978). On the ecology of the Asian elephant, *Elephas maximus* Linn., with particular reference to Malaya and Sri Lanka. PhD thesis, University of Cambridge.

Orians, G. H. (1969). The number of bird species in some tropical forests. *Ecology*, **50**, 783–801.

Orians, G. H. (1975). Diversity, stability and maturity in natural ecosystems. In *Unifying concepts in ecology*, ed. W. H. van Dobben & R. H. Lowe McConnell, pp. 139–50. The Hague: Junk.

Palmer, J. & Synnott, T. (1992). The management of natural forests. In *Managing the world's forests*, ed. N. P. Sharma, pp. 337–73. Dubuque, Iowa: Kendall/Hunt Publishing.

Panayotou, T. & Ashton, P. S. (1992). *Not by timber alone: economics and ecology for sustaining tropical forests*. Washington DC: Island Press.

Parsons, M. J. (1983). A conservation study of the birdwing butterflies *Ornithoptera* and *Triodes* (Lepidoptera: Papilionidae) in Papua New Guinea. Report to the Department of Primary Industry, Bulolo, Papua New Guinea.

Payne, J. (1988). *Orang-utan conservation in Sabah*. Kuala Lumpur: WWF-Malaysia.

Peters, C. M., Gentry, A. H. & Mendelsohn, R. O. (1989). Valuation of an Amazonian rain forest. *Nature*, **339**, 655–6.

Petranka, J. W., Brannon, M. P., Hopey, M. E. & Smith, C. K. (1994). Effects of timber harvesting on low elevation populations of southern Appalachian salamanders. *Forest Ecology and Management*, **67**, 135–47.

Petter, J-J., & Peyrièras, A. (1974). A study of population density and home ranges of *Indri indri* in Madagascar. In *Prosimian Biology*, ed. R. D. Martin, G. A. Doyle & A. C. Walker, pp. 39–48. London: Duckworth.

Pimm, S. L. (1980). Food web design and the effect of species deletions. *Oikos*, **35**, 139–49.

Pimm, S. L. & Lawton, J. H. (1978). On feeding from more than one trophic level. *Nature*, **275**, 542–4.

Plumptre, A. J. (1995). The Budongo Forest Reserve: the effects of 60 years of selective logging on tree distributions and forest structure. *Commonwealth Forestry Review*, **74**, 253–8.

Plumptre, A. J. & Reynolds, V. (1994). The effect of selective logging on the primate populations in the Budongo Forest reserve, Uganda. *Journal of Applied Ecology*, **31**, 631–41.

Plumptre, A. J., Reynolds, V. & Bakuneeta, C. (1994). The contribution of fruit eating primates to seed dispersal and natural regeneration after selective logging. Final report of ODA project R4738, Institute of Biological Anthropology, Oxford.

Poels, R. H. L. (1987). *Soils, water and nutrients in a forest ecosystem in Suriname.* Wageningen, The Netherlands: Pudoc.

Poore, D. & Sayer, J. (1987). *The management of tropical moist forest lands: ecological guidelines.* Gland, Switzerland: IUCN.

Poore, D., Burgess, P., Palmer, J., Reitbergen, S. & Synnott, T. (1989). *No timber without trees: sustainability in the tropical forest.* London: Earthscan Publications.

Potter, G. L. (1975). Possible climatic impacts of tropical deforestation. *Nature*, **258**, 697–8.

Proctor, J. (1992). Soils and mineral nutrients: what do we know, and what do we need to know, for wise rain forest management. In *Wise management of tropical forests*, ed. F. R. Miller & K. L. Adam, pp. 27–35. Oxford: Oxford Forestry Institute.

Proud, K. R. S. & Hutchinson, I. D. (1980). Management of natural reserves to maintain faunal diversity: a potential use for forest silviculture. In *Tropical ecology and development*, vol. 1, ed. J. I. Furtado, pp. 247–55. Kuala Lumpur, Malaysia: International Society for Tropical Ecology.

Putman, R. J. (1994). *Community ecology.* London: Chapman and Hall.

Queensland Department of Forestry (1987). *Aspects of management in plantations in tropical and sub-tropical Queensland.* Brisbane: Department of Forestry.

Raiche, J. W. & Christensen, N. C. (1989). Malaysian dipterocarp forest: tree seedling and sapling species composition and small-scale disturbance patterns. *National Geographic Research*, **5**, 348–63.

Raiche, J. W. & Gong, W. K. (1990). Effects of canopy opening on tree seed germination in a Malaysian dipterocarp forest. *Journal of Tropical Ecology*, **6**, 203–17.

Ralph, C. J. (1985). Habitat association patterns of forest and steppe birds of northern Patagonia, Argentina. *Condor*, **87**, 471–83.

Ramirez, M. F., Freese, C. H. & Revilla, J. (1977). Feeding ecology of the pygmy marmoset *Cebuella pygmaea* in north-eastern Peru. In *The biology and conservation of the Callitrichidae*, ed. D. G. Kleiman, pp. 91–104. Washington, DC: Smithsonian Institution Press.

Rapera, R. B. (1978). Effects of logging on residual stands. *BIOTROP Special Publications*, **3**, 119–25.

Recher, H. F., Shields, J., Kavanagh, R. P. & Webb, G. (1987). Retaining remnant mature forest for nature conservation at Eden, New South Wales: a review of theory and practice. In *Nature conservation: the role of remnants of vegetation*, ed. D. A. Saunders, G. W. Arnold, A. A. Burbidge & A. J. Hopkins, pp. 177–94. Chipping Norton, Australia: Surrey Beaton and Sons.

Redhead, J. F. (1960). An analysis of logging damage in lowland rain forest, western Nigeria. *Nigerian Forest Information Bulletin (New Series)*, **10**, 5–16.

Reid, W. V. (1992). How many species will there be? In *Tropical deforestation and species extinction*, ed. T. C. Whitmore and J. A. Sayer, pp. 55–73. London: Chapman and Hall.

Remsen, J. V. Jr & Parker, T. A. III. (1984). Contribution of river-created habitats to bird species richness in Amazonia. *Biotropica*, **15**, 223–31.

Repetto, R. (1988). Overview. In *Public policies and the misuse of forest resources*, ed. R. Repetto & M. Gillis, pp. 1–41. Cambridge: Cambridge University Press.

Richardson, S. D. (1970). The end of forestry in Great Britain. *Commonwealth Forestry Review*, **49**, 324–35.

Robinson, J. G. (1993). The limits of caring: sustainable living and the loss of biodiversity. *Conservation Biology*, **7**, 20–8.

Robinson, M. H. (1969). The defensive behaviour of some orthopteroid insects from Panama. *Transactions of the Royal Entomological Society of London*, **121**, 281–303.

Roche, L. & Dourojeanni, M. J. (1984). *A guide to in situ conservation of genetic resources in tropical woody species*. Rome: FAO.

Royal Society (1987). Soil fauna in primary and logged-over forest in Sabah. *South-east Asian Rain Forest Research Programme Newsletter*, **3**, 4–5.

Ruitenbeck, H. J. (1988). *Social cost-benefit analysis of the Korup Project, Cameroon*. Godalming, UK: WWF-UK.

Rukuba, M. L. S. B. (1992). Linkages between wise forest management for forest production and local markets and industries: example from Uganda. In *Wise management of tropical forests*, ed. F. R. Miller & K. L. Adam, pp. 161–9. Oxford: Oxford Forestry Institute.

Salleh, M. N. & Baharudin, J. (1985). Silvicultural practices in peninsular Malaysia. In *The future of tropical rain forests in South-east Asia*. Gland, Switzerland: IUCN Commission on Ecology Paper no. 10, 81–4.

Samat, A. (1993). The effects of rain forest disturbance on fish populations in Sabah, Malaysia. MSc thesis, University of Aberdeen.

Sandström, U. (1992). Cavities in trees: their occurrence, formation and importance for hole-nesting birds in relation to silvicultural practise. PhD thesis, Swedish University of Agricultural Sciences, Uppsala.

Saunders, D. A. & de Rebeira, C. P. (1991). Values of corridors to avian populations in a fragmented landscape. In *Nature conservation, 2. The role of corridors*, ed. D. A. Saunders & R. J. Hobbs, pp. 221–40. Chipping Norton, Australia: Surrey Beaton and Sons.

Saunders, D. A. & Hobbs, R. J., ed. (1991). *Nature conservation, 2. The role of corridors*. Chipping Norton, Australia: Surrey Beaton and Sons.

Sayer, J. A. (1991). *Rainforest buffer zones: guidelines for protected area managers*. Gland, Switzerland: IUCN.

Sayer, J. A., Harcourt, C. & Collins, M. N. (1992). *The conservation atlas of tropical forests: Africa*. London: Macmillan.

Sayer, J. A. & Whitmore, T. C. (1991). Tropical moist forests: destruction and species extinction. *Biological Conservation*, **55**, 199–214.

Schreiber, B. & deCalestra, D. S. (1992). The relationship between cavity-nesting birds and snags on clearcuts in western Oregon. *Forest Ecology and Management*, **50**, 299–316.

Schultz, J. P. (1960). *Ecological studies on rain forest in northern Suriname*. Amsterdam: North-Holland.

Sedjo, R. A. (1989). Forests: a tool to moderate global warming? *Environment*, **31**, 14–20.

Shepherd, K. R. & Richter, H. V. (1985). *Managing the tropical forest*. Canberra: Australian National University Press.

Shuttleworth, W. J. and Nobre, C. A. (1992). Wise forest management and its linkages to climate change. In *Wise management of tropical forests*, ed. F. R. Miller and K. L. Adam, pp. 77–90. Oxford: Oxford Forestry Institute.

Silva, J. M. N., de Carvalho, J. O. P., Lopes, J. da C. A., de Almeida, B. F., Costa, D. H. M., de Oliveira, L. C., Vanclay, J. K. & Skovsgaard, J. P. (1995). Growth and yield of a tropical rain forest in the Brazilian Amazon 13 years after logging. *Forest Ecology and Management*, **71**, 267–74.

Simberloff, D. (1986). Design of nature reserves. In *Wildlife conservation evaluation*, ed. M. B. Usher, pp. 315–37. London: Chapman and Hall.

Simberloff, D. & Cox, J. (1987). Consequences and costs of conservation corridors. *Conservation Biology*, **1**, 63–71.

Skorupa, J. P. (1986). Responses of rain forest primates to selective logging in the Kibale Forest, Uganda: a summary report. In *Primates: the road to self-sustaining populations*, ed. K. Benirschke, pp. 57–70. Berlin, Heidelberg, New York: Springer.

Skorupa, J. P. (1988). The effects of selective timber harvesting on rain forest primates in Kibale Forest, Uganda. PhD thesis, University of California, Davis.

Skorupa, J. P. & Kasenene, J. M. (1984). Tropical forest management: can rates of natural treefalls help guide us? *Oryx*, **18**, 96–101.

Smits, W. (1985). Dipterocarp mycorrhizae. In *The future of tropical rain forests in South-east Asia*. Gland, Switzerland: IUCN Commission on Ecology Paper, **10**, 51–4.

Soini, P. (1982). Ecology and population dynamics of the pygmy marmoset *Cebuella pygmaea*. *Folia Primatologica*, **39**, 1–21.

Soong, N. K., Haridas, G., Yeoh, C. S. & Tan, P. H. (1980). *Soil erosion and conservation in Malaysia*. Kuala Lumpur: Rubber Research Institute of Malaysia.

Soulé, M. E. ed. (1987). *Viable populations for conservation*. Cambridge: Cambridge University Press.

Stork, N. E. (1988). Insect diversity: facts, fiction and speculation. *Biological Journal of the Linnaean Society*, **35**, 321–7.

Struhsaker, T. T. (1987). Forestry issues and conservation in Uganda. *Biological Conservation*, **39**, 209–34.

Struhsaker, T. T., Kasenene, J. M., Gaither, J. C. Jr, Larsen, N., Musango, S. & Bancroft, R. (1989). Tree mortality in the Kibale Forest , Uganda: a case of dieback in a tropical rain forest adjacent to exotic conifer plantations. *Forest Ecology and Management*, **29**, 165–85.

Sutton, S. L. & Collins, N. M. (1991). Insects and tropical forest conservation. In *The conservation of insects and their habitats*, ed. N. M. Collins & J. A. Thomas, pp. 405–24. London: Academic Press.

Synnott, T. J. (1979). *A manual of permanent plot procedures for tropical rain forest*. Oxford: Commonwealth Forestry Institute.

Synnott, T. J. (1992). The introduction of basic management into tropical forests. In *Wise management of tropical forests*, ed. F. R. Miller & K. L. Adam, pp. 91–5. Oxford: Oxford Forestry Institute.

Tabor, G. M., Johns, A. D. and Kasenene, J. M. (1990). Deciding the future of Uganda's tropical forests. *Oryx*, **24**, 208–14.

Tang, H. T., Haron, H. A. H. & Cheah, E. K. (1981). Mangrove forests of peninsular Malaysia: a review of management and research objectives and priorities. *Malaysian Forester*, **44**, 77–92.

Taylor, R. J. & Haseler, M. E. (1995). Effects of partial logging systems on bird assemblages in Tasmania. *Forest Ecology and Management*, **72**, 131–49.

Terborgh, J. (1983). *Five New World primates*. Princeton, New Jersey: Princeton University Press.

Terborgh, J. (1992). Maintenance of diversity in tropical forests. *Biotropica*, **24**, 283–92.

Terborgh, J., Robinson, S. K., Parker, T. A. III, Munn, C. A. & Pierpont, N. (1990). Structure and organization of an Amazonian bird community. *Ecological Monographs*, **60**, 213–38.

Terborgh, J. & Weske, J. S. (1969). Colonization of secondary habitats by Peruvian birds.

*Ecology*, **50**, 765–82.

Terborgh, J. & Winter, B. (1980). Some causes of extinction. In *Conservation biology*, ed. M. E. Soulé & B. A. Wilcox, pp. 119–33. Sunderland, Mass.: Sinauer.

Thiollay, J-M. (1992). Influence of selective logging on bird species diversity in a Guianan rain forest. *Conservation Biology*, **6**, 47–63.

Thomas, J. W., Forsman, E. D., Lint, J. B., Meslow, E. C., Noon, B. R. & Verner, J. (1990). *A conservation strategy for the northern spotted owl*. Portland, Oregon: USDA Forest Service.

Thomas, S. C. (1991). Population densities and patterns of habitat use among anthropoid primates of the Ituri Forest, Zaire. *Biotropica*, **23**, 68–83.

Thompson, W. A., Stocker, G. C. & Kriedermann, P. E. (1988). Growth and photosynthetic response to light and nutrients of *Flindersia brayleyana* F. Muell., a rainforest tree with broad tolerance to sun and shade. *Australian Journal of Plant Physiology*, **15**, 299–315.

Tinal, U. & Palenewen, J. L. (1978). Mechanical logging damage after selective logging in the lowland dipterocarp forest at Baloro, East Kalimantan. *BIOTROP Special Publication*, **3**, 91–6.

Uhl, C. & Buschbacher, R. (1985). A disturbing synergism between cattle ranch burning practices and selective tree harvesting in the eastern Amazon. *Biotropica*, **17**, 265–8.

UK Government (1994). *Sustainable forestry: the UK programme*. London: HMSO.

UNDP/FAO (1983). Forestry development Brazil: project findings and recommendations. UNDP/FAO Field Document BRA/78/003, terminal report.

UNEP (1995). *Global biodiversity assessment*. Cambridge: Cambridge University Press.

Veríssimo, A., Barreto, P., Tarifa, R. & Uhl, C. (1995). Extraction of a high-value natural resource in Amazonia: the case of mahogany. *Forest Ecology and Management*, **72**, 39–60.

Vincent, J., Gandapur, A. K. & Brooks, D. J. (1990). Species substitution and tropical log imports by Japan. *Forest Science*, **36**, 657–64.

Vincent, J., Brooks, D., Gandapur, A. K. & Alamgir, K. (1991). Substitution between tropical and temperate logs. *Forest Science*, **37**, 1484–91.

Wells, D. R., Hails, C. J. & Hails, A. J. (1978). A study of the birds of Gunung Mulu National Park, Sarawak, with special emphasis on those of lowland forests. Report to the Royal Geographical Society, London.

Welsh, C. J. E. & Copen, D. E. (1992). Availability of nesting sites as a limit to woodpecker populations. *Forest Ecology and Management*, **48**, 31–41.

White, L. J. T. (1992). Vegetation history and logging disturbance: effects on rain forest mammals in the Lope Reserve, Gabon (with special emphasis on elephants and apes). PhD thesis, University of Edinburgh.

White, L. J. T. & Tutin, C. E. G. (in press). Why chimpanzees and gorillas respond differently to logging: a case study from Gabon that cautions against generalizations. In *African rainforest ecology and conservation*, ed. B. Weber, A. Veder, H. Simons-Morland, L. White & T. Hart. New Haven: Yale University Press.

Whitmore, T. C. (1984). *Tropical rain forests of the Far East*, 2nd edn. Oxford: Clarendon Press.

Whitmore, T. C. (1990). *An introduction to tropical rain forests*. Oxford: Oxford University Press.

Whitmore, T. C. (1991). Tropical rain forest dynamics and its implications for management. In *Rain forest regeneration and management*, ed. A. Gómez-Pompa, T. C. Whitmore & M. Hadley, pp. 67–89. Paris: UNESCO.

Whitmore, T. C. & Sayer, J. A. (1992). Deforestation and species extinction in tropical moist forest. In *Tropical deforestation and species extinction*, ed. T. C. Whitmore & J. A. Sayer, pp. 1–14. London: Chapman and Hall.

Whitmore, T. C. & Silva, J. N. M. (1990). Brazil rain forest timbers are mostly very dense. *Commonwealth Forestry Review*, **69**, 87–90.

Whitten, A. J. & Damanik, S. J. (1986). Mass defoliation of mangroves in Sumatra, Indonesia. *Biotropica*, **18**, 176.

Whitten, A. J., Damanik, S. J., Anwar, J. & Hisyam, N. (1984). *The ecology of Sumatra*. Yogyakarta, Indonesia: Gadjah Mada University Press.

Wiersum, K. F. (1982). Tree gardening and taungya on Java: examples of agroforestry techniques in the humid tropics. *Agroforestry Systems*, **1**, 53–70.

Willis, E. O. (1974). Populations and local extinctions of birds on Barro Colorado Island, Panama. *Ecological Monographs*, **44**, 153–69.

Willis, E. O. (1979). The composition of avian communities in remanescent woodlots in southern Brazil. *Papèis Avulsos de Zoologia, Museu de Sao Paulo*, **33**, 1–25.

Willson, M. F. (1974). Avian community organization and habitat structure. *Ecology*, **55**, 1017–29.

Wilson, E. O. (1993). Forest ecosystems: more complex than we know. In *Defining sustainable forestry*, ed. G. H. Aplet, N. Johnson, T. Olson & V. Alaric Sample, pp. xi–xiii. Washington, DC: Island Press.

Wilson, W. L. & Johns, A. D. (1982). Diversity and abundance of selected animal species in undisturbed forest, selectively logged forest and plantations in East Kalimantan, Indonesia. *Biological Conservation*, **24**, 205–18.

Wong, M. (1982). Patterns of food availability and understory bird community structure in a Malaysian rain forest. PhD thesis, University of Michigan, Ann Arbor.

Wong, M. (1985). Understory birds as indicators of regeneration in a patch of selectively logged West Malaysian rain forest. In *Conservation of tropical forest birds*, ed. A. W. Diamond & T. E. Lovejoy, pp. 249–63. Cambridge: International Council for Bird Preservation.

World Bank (1991). *Forest policy paper*. Washington, DC: The World Bank.

WRI (1985). *Tropical forests: a call for action*. Washington, DC: World Resources Institute.

WRI (1994). *World Resources: 1994–5*. New York: Oxford University Press.

WWF (1991). *Tropical rain forests and the environment: a survey of public attitudes*. Godalming, UK: WWF-UK.

Wyatt-Smith, J. (1963). Manual of Malayan silviculture for inland forests, 2 vols. *Malayan Forest Records*, 23.

Wyatt-Smith, J. (1987). The management of tropical moist forest for the sustained production of timber: some issues. *IUCN/IIED Tropical Forest Policy Paper*, 4.

Zadroga, F. (1981). The hydrological importance of a montane cloud forest area of Costa Rica. In *Tropical agricultural hydrology*, ed. R. Lal & F. W. Russell, pp. 59–73. New York: Wiley.

Zobel, B. J., van Wyk, G. & Stahl, P. (1987). *Growing exotic forests*. New York: Wiley Interscience.

# SUBJECT INDEX